MEMORY-BASED LANGUAGE PROCESSING
基于记忆的语言处理

[比] Walter Daelemans
[荷] Antal van den Bosch 著

孙　栩　导读

北京大学出版社
PEKING UNIVERSITY PRESS

著作权合同登记号 图字：01-2014-3772

图书在版编目(CIP)数据

基于记忆的语言处理＝Memory-Based Language Processing：英文/(比)戴勒曼斯(Daelemans, W.)，(荷)博施(Bosch, A.V.D.)著. —北京：北京大学出版社，2015.7
(计算语言学与语言科技原文丛书)
ISBN 978-7-301-25909-2

Ⅰ.①基… Ⅱ.①戴…②博… Ⅲ.①自然语言处理—研究—英文 Ⅳ.①TP391

中国版本图书馆CIP数据核字(2015)第117666号

Memory-Based Language Processing, first edition (ISBN 978-0-521-11445-5) by Walter Daelemans and Antal van den Bosch first published by Cambridge University Press 2005
All rights reserved.
This reprint edition for the People's Republic of China is published by arrangement with the Press Syndicate of the University of Cambridge, Cambridge, United Kingdom.
© Cambridge University Press & Peking University Press 2015
This book is in copyright. No reproduction of any part may take place without the written permission of Cambridge University Press and Peking University Press.
This edition is for sale in the People's Republic of China (excluding Hong Kong SAR, Macau SAR and Taiwan Province) only.
此版本仅限在中华人民共和国(不包括香港、澳门特别行政区及台湾地区)销售。

书　　名	基于记忆的语言处理
著作责任者	[比] Walter Daelemans　[荷] Antal van den Bosch　著
责任编辑	李凌
标准书号	ISBN 978-7-301-25909-2
出版发行	北京大学出版社
地　　址	北京市海淀区成府路205号　100871
网　　址	http://www.pup.cn　新浪微博:@北京大学出版社
电子信箱	zpup@pup.cn
电　　话	邮购部 62752015　发行部 62750672　编辑部 62753374
印刷者	北京大学印刷厂
经销者	新华书店
	787毫米×980毫米　16开本　13.25印张　245千字
	2015年7月第1版　2015年7月第1次印刷
定　　价	32.00元

未经许可，不得以任何方式复制或抄袭本书之部分或全部内容。
版权所有，侵权必究
举报电话：010-62752024　电子信箱：fd@pup.pku.edu.cn
图书如有印装质量问题，请与出版部联系，电话：010-62756370

"计算语言学与语言科技原文丛书"由北京大学—香港理工大学汉语语言学研究中心、北京大学计算语言学研究所(由973课题"文本内容理解的数据基础"、863课题"大规模汉语语义基础资源库和知识库设计构建及工具平台"支持)和北京大学出版社合作推出

学术委员会
Academic Advisory Committee

主 任：
　　黄居仁（香港）

委 员：
　　Chris Manning（Stanford）　　　Harold Somers（Dublin）
　　Maarten de Rijke（Amsterdam）　Suzanne Stevenson（Toronto）
　　陈克健（台北）　　　　　　　　冯志伟（北京）
　　李宇明（北京）　　　　　　　　陆俭明（北京）
　　郭　锐（北京）　　　　　　　　石定栩（香港）
　　苏克毅（台北）　　　　　　　　孙茂松（北京）
　　王厚峰（北京）　　　　　　　　王士元（香港）
　　俞士汶（北京）　　　　　　　　松木裕治（奈良）
　　郑锦全（Urbana-Champaign）　　邹嘉彦（香港）

编委会
Editorial Committee

主　编：
　　黄居仁教授(香港)

编　委：
　　顾日国教授(北京)　　　　黄萱菁教授(上海)
　　姬东鸿教授(武汉)　　　　刘　群教授(Dublin)
　　陆　勤教授(香港)　　　　蒙美玲教授(香港)
　　苏新春教授(厦门)　　　　孙　栩研究员(北京)
　　夏　飞教授(Seattle)　　　徐飞玉教授(Saarbrücken)
　　薛念文教授(Waltham)　　　曾淑娟副研究员(台北)
　　詹卫东教授(北京)　　　　张凤珠编审(北京)
　　赵铁军教授(哈尔滨)　　　周　明研究员(北京)
　　宗成庆研究员(北京)　　　常宝宝副教授(执行秘书)(北京)

丛书前言

"计算语言学与语言科技原文丛书"于2010年创立，2010 COLING 国际计算语言学会议在北京举办之前出版了第一批图书。本丛书的出版象征着国内计算语言学研究与国际的接轨。国内学者正跻身计算语言学的国际舞台：一些资深学者已在 COLING 两个最主要的国际会议/组织中获选并担任重要的领导职务；而积极参与这些重要的国际会议也已在年轻学者中蔚然成风，他们已可谓是会议的主流参与者之一。在这样的氛围中，希望本丛书第二批图书的出版，能让国内有心投入语言科技与计算语言学研究的学者们如虎添翼，在国际舞台上创新并引导议题！

计算语言学(Computational Linguistics, CL)在语言科学与信息科学的研究中扮演着关键性的角色。语言学理论寻求对语言现象进行规律性的预测，做出完整的解释，计算语言学正好为这两点提供了验证与应用的大好机会。作为语言学、信息科学乃至于心理学与认知科学结合的交叉学科，计算语言学更为语言学基础研究与福国利民应用研究的接轨提供了绝佳界面。事实上，计算语言学与人类语言科技(Human Language Technology, HLT)可以视为体用两面，不可切分。

计算语言学研究的滥觞，其实源于上世纪五六十年代的机器翻译研究，中文计算语言学的研究也几乎同步开始。在美国伯克利加州大学研究室，王士元、邹嘉彦、C.Y. Dougherty 等人1960年已开始研究中英、中俄机器翻译。他们的研究是与世界最尖端的科技同步的。国内中俄翻译研究也不遑多让，大约在20世纪50年代中期便已开始。可惜的是，这些中文方面早期机器翻译研究，由于硬件与软件的限制，未能有效传承下来。中文计算语言学研究比较系统的发展始于1986年。这一年，海峡两岸不约而同地分别成立了两个致力于建立中文计算语言学基础架构的研究群：北京大学的计算语言学研究所，在朱德熙先生倡议下成立，随后一段时间由陆俭明、俞士汶主持；而台北"中研院"的中文词知识库小组，由谢清俊创立，陈克健主持，黄居仁1987年回去后加入。

中文计算语言学的研究，近30年来已累积了相当可观的成绩。计算语言学的重要研究领域与议题中都能看到中文方面的相关研究成果，华人计算语言学学者也渐渐在国际学术界崭露头角。随着世界经济转向知识密集型产业,跨语言跨文化沟通与知识整合成为知识产业的关键环节,语言科技的发展成为国际主流指日可待。在这个有利发展的大环境下,我们期待看到,中文计算语言学与华人计算语言学学者的成绩,百尺竿头更进一步,中文方面的研究可以进入计算语言学的学术核心,能够产生有能力引导议题并掌控研究方向的大师。

回顾国内的计算语言学发展,计算机科学的贡献多于语言学的贡献。这个现象,在理论与概率模型整合研究的趋势中,不免令人忧心。语言学的贡献弱,或许可以部分归咎于英文研究专著在国内不易取得;而比较容易取得的期刊或会议论文,在篇幅的限制下,又往往无法对理论做深入完整的铺陈,从而导致国内的年轻学者长于运算而拙于理据。因此,在期待大师与引领世界研究潮流两个方向,藉由英文专书来巩固研究理据,进而开拓研究视野,是非常重要的一步。

"计算语言学与语言科技原文丛书"的引进,就是在上述背景下促成的。个人忝为剑桥大学出版社"自然语言处理研究"(Studies in Natural Language Processing, SNLP)系列的主编,对于将此系列中较重要的几本书引入国内,责无旁贷。第二批出版的原文书,除了剑桥大学出版社的图书外,还有施普林格出版公司(Springer)语言科技系列中的几本书,以进一步拓展领域涵盖面。引进原书,原样出版,是容易的,然而若要真正搭建知识的桥梁,使国内学者与学生不仅能开拓研究视野,更能将原文著作的理论精髓应用于中文研究,则实在不易。因此,本系列每本书我们都邀请了一位专家撰写中文导读。这些导读可以说是本系列的精华、重点,使每本书比剑桥和施普林格的原本增加了不少附加价值。

每篇中文导读都包括三个重要的组成部分。第一部分是全书内容概要的介绍。导读专家都是长年浸淫于该领域的学者,他们能提纲挈领,并提供相关研究背景。因此,通过阅读导读,读者更易掌握并吸收该书的重要内容。第二部分是中文相关研究。原文著作不见得会提到相关的中文研究,由导读专家补充介绍,搭起理论与中文相关应用的桥梁,更能引导读者找到在这个议题进入中文研究的最佳切入点,让中文相关研究的开拓者的成绩更能发扬光大。第三部分重点在于补充原书出版后该领域研究的新发展。现代科技发展迅速,任何经典著作出版后,几乎马上就有新的相关研究。因此,在理论架构的脉络中,加上新近的发展,能使读者更贴切地掌握研究脉动。全书的内容摘要通常以文字叙述,而中文相关研究及最新研究发展则分别以文字叙述和延伸阅读书目的方式呈现。延伸阅读书目,可以使读者很快上手,进入相关研究领域,也是本系列的重要设计之一。

本丛书2010年出版第一批图书,现在出版第二批图书,必须感谢许多同行的付出。在规划出版的漫长过程中,北大计算语言学研究所的俞士汶老师及常宝宝老师一直无私无悔地支持。而香港理工大学的挹注,北大—理大汉语语言学研究中心石定栩、郭锐等几位的支持,使得整个系列能够顺利出版。此外,还要感谢北京大学出版社王飙主任、杜若明编审及李凌编辑,他们认同我们的宗旨,落实了丛书的出版工作。最后,感谢丛书的国内编委,特别是此次担任导读主笔的各位,正是他们脑力与心血的付出,才替读者们搭建了进入学术殿堂的平台。

丛书主编 黄居仁
谨志于香港,红磡
2014年9月

目 录

导读 1

Preface   **1**

1 Memory-Based Learning in Natural Language Processing   **3**
   1.1 Natural language processing as classification 6
   1.2 A linguistic example 9
   1.3 Roadmap and software 12
   1.4 Further reading 14

2 Inspirations from linguistics and artificial intelligence   **15**
   2.1 Inspirations from linguistics 15
   2.2 Inspirations from artificial intelligence 21
   2.3 Memory-based language processing literature 22
   2.4 Conclusion 24

3 Memory and Similarity   **26**
   3.1 German plural formation 27
   3.2 Similarity metric 28
      3.2.1 Information-theoretic feature weighting 29
      3.2.2 Alternative feature weighting methods 31
      3.2.3 Getting started with TiMBL 32
      3.2.4 Feature weighting in TiMBL 36
      3.2.5 Modified value difference metric 38
      3.2.6 Value clustering in TiMBL 39
      3.2.7 Distance-weighted class voting 42
      3.2.8 Distance-weighted class voting in TiMBL 44
   3.3 Analyzing the output of MBLP 45
      3.3.1 Displaying nearest neighbors in TiMBL 45
   3.4 Implementation issues 46
      3.4.1 TiMBL trees 47

3.5	Methodology		47
	3.5.1 Experimental methodology in TIMBL		48
	3.5.2 Additional performance measures in TIMBL		52
3.6	Conclusion		55

4 Application to morpho-phonology — 57

4.1	Phonemization		59
	4.1.1 Memory-based word phonemization		59
	4.1.2 TREETALK		60
	4.1.3 IGTREE in TIMBL		67
	4.1.4 Experiments: applying IGTREE to word phonemization		69
	4.1.5 TRIBL: trading memory for speed		71
	4.1.6 TRIBL in TIMBL		73
4.2	Morphological analysis		73
	4.2.1 Dutch morphology		74
	4.2.2 Feature and class encoding		74
	4.2.3 Experiments: MBMA on Dutch wordforms		76
4.3	Conclusion		80
4.4	Further reading		83

5 Application to shallow parsing — 85

5.1	Part-of-speech tagging		86
	5.1.1 Memory-based tagger architecture		87
	5.1.2 Results		88
	5.1.3 Memory-based tagging with MBT and MBTG		90
5.2	Constituent chunking		96
	5.2.1 Results		96
	5.2.2 Using MBT and MBTG for chunking		97
5.3	Relation finding		99
	5.3.1 Relation finder architecture		99
	5.3.2 Results		100
5.4	Conclusion		101
5.5	Further reading		102

6 Abstraction and generalization — 104

6.1	Lazy versus eager learning		106
	6.1.1 Benchmark language learning tasks		107
	6.1.2 Forgetting by rule induction is harmful in language learning		111
6.2	Editing examples		115

	6.3	Why forgetting examples can be harmful	123
	6.4	Generalizing examples	128
		6.4.1 Careful abstraction in memory-based learning	128
		6.4.2 Getting started with FAMBL	135
		6.4.3 Experiments with FAMBL	137
	6.5	Conclusion	143
	6.6	Further reading	145
7	**Extensions**		**148**
	7.1	Wrapped progressive sampling	149
		7.1.1 The wrapped progressive sampling algorithm	150
		7.1.2 Getting started with wrapped progressive sampling	152
		7.1.3 Wrapped progressive sampling results	154
	7.2	Optimizing output sequences	156
		7.2.1 Stacking	157
		7.2.2 Predicting class n-grams	160
		7.2.3 Combining stacking and class n-grams	162
		7.2.4 Summary	164
	7.3	Conclusion	164
	7.4	Further reading	165
Bibliography			**168**
Index			**186**

导　读

孙　栩

1　学科背景和内容提要

　　自然语言处理包括逻辑主义流派和经验主义流派。从20世纪90年代开始，经验主义流派在自然语言处理领域占据了主流。原因之一在于数据的大规模化，使得经验主义的方法能够获得大规模数据的支持。本书介绍的"基于记忆的语言处理"根植于经验主义流派。基于记忆的语言处理，简而言之，就是使用"基于记忆的学习"这种特殊的机器学习方法来处理语言。基于记忆的学习来源于这样一种假设或者说启发：当人们对一件事情进行分析判断的时候，往往会直接参考最相似的经验或者记忆，而不是从这些经验或者记忆中抽象出规则来进行判断。

　　本书作者强调，一系列的实验表明，基于记忆的自然语言处理在多个任务上获得了很好的效果，并初步解释了基于记忆的语言处理能够取得好效果的原因——语言处理数据常常含有大量互相冲突的数据模式，而且往往伴随着数据稀疏的特点。换句话说，很多语言现象不常见，而且互相冲突。现有的大部分方法是"急切(eager)"的方法，这些方法总是试图从数据中抽象出规则或者简单的参数模型，并且过滤掉冲突的、不常见的数据模式。这类急切方法难以在语言处理任务上取得好效果，因为这类数据模式冲突、低频率的语言现象其实不是噪音，而是重要的语言现象，所以不能在抽象、泛化过程中过滤掉这些大量而特殊的语言现象。基于记忆的学习技术恰恰解决了这个问题——基于记忆的学习是一种"懒惰(lazy)"的学习方法，因为其学习过程非常简单，就是简单地把所有的训练数据保存在"记忆"中，从而不存在急切方法所具有的抽象(abstraction)问题。

　　本书主要面向自然语言处理／计算语言学和机器学习方面的学习者。对于机器学习研究来说，语言处理提供了广泛的、具有挑战性的现实任务和数据，比如海量的特征(features)、超过百万量级的大规模数据、复杂的数据结构等。对于自然语言处理／计算语言学方面的学者，基于记忆的学习提供了一个独特的符号化统计学习工具用于处理语言。总体来说,本书浅显易懂,即使是刚接触这方面内容的学习者,也能轻松读懂本书。

本书简单实用,介绍了一个直观又实用的方法——基于记忆的学习。跟很多现有的自然语言处理技术相比,本书介绍的"基于记忆的学习"包括两个主要方面:一是学习方面,即简单地存贮记忆,也就是训练实例;二是遇到新实例需要分类时,就简单地找到最相似的记忆/实例,然后进行适当的修改,从而完成分类。本书另有配套的软件发布。在介绍了核心方法之后,又详细说明作者和其团队开发的基于记忆学习的软件包 TIMBL,更凸显了实用性。

2 内容介绍

第一章:基于记忆学习的自然语言处理

本章对背景知识、术语进行了介绍,包括自然语言处理的目标、历史、流派。对语言处理的方式——一般是在本质上降解为分类问题——进行了介绍。最主要的,是对"基于记忆的语言处理(memory-based language processing, MBLP)"以及"基于记忆的学习(memory-based learning, MBL)"做了浅显易懂的介绍。通过一些比较直观的例子,本章对"基于记忆的学习"给出了一个初步的定义——基于记忆的学习包括两个主要方面:一是学习方面,即简单地存贮记忆和经验;二是解决新实例的分类问题,即简单地找到最相似的记忆,然后进行适当的修改。

第二章:来自语言学和人工智能的启发

本章阐述了基于记忆的语言处理技术的思想起源,包括语言学、人工智能、认知语言学。简而言之,"基于记忆的语言处理"就是使用"基于记忆的学习"技术处理语言;而"基于记忆的学习"就是把所有的训练数据都存储在记忆里——对于计算机来说,即内存等存储设备——然后当需要对新实例进行分类的时候,就搜索记忆中最相似的实例作为参考进行分类。这个主意并不是全新的,而是来源于更早的科研探索,包括语言学、认知科学、人工智能等。在人工智能的相关研究里面,有跟"基于记忆"的方法类似的思路和方法。典型代表有 20 世纪 50 年代就开始发展的"最近邻分类方法(nearest-neighbor classifier)"。Fix 和 Hodges 在其 1952 年的著作中提到,"最近邻"这个思想在很多现实场合都很符合直觉,比如,医生对某个病症开处方的时候往往会参考以前对类似病症所开的处方。

但是,原始的最近邻分类方法——包括最具代表性的k-NN方法——在被提出很长一段时间内应用和影响都比较有限。这主要是因为初期的k-NN方法在存储和计算方面有若干不足——存储的时候需要把所有的实例存储在内存中,导致了内存代价较高;此外,对新的实例进行分类计算的时候,需要跟存储的所有实例进行比较,导致了计算复杂度较高。从20世纪80年代开始,由于最近邻方法简单、符合直觉的优点,其在人工智能的多个领域重新获得重视,并且发展出了多个变种,包括 memory-based reasoning、case-based reasoning、exemplar-based learning、locally-weighted learning,以及 instance-based learning 等。这些方法在不同的方面扩展了初期的k-NN方法,以解决上面提到的k-NN缺点。本章还简要介绍了和基于记忆的语言处理技术密切相关的其他技术,包括"基于实例的机器翻译(example-based machine translation, EBMT)"以及"面向数据的句法分析(data-oriented parsing)"。

第三章:记忆和相似度

本章介绍了对基于记忆的自然语言处理技术的具体实现。一个重要的计算步骤是相似度(similarity)的计算,章节3.2对相似度计算的实现和数学表达式进行了介绍。相似度计算有不同的版本,最简单的情况是只考虑特征向量(feature vector)的相似度,但是这样计算出来的相似度往往不够准确,因为没有考虑特征的权重——也就是说,有的特征重要,有的特征不重要,应该在计算相似度的过程中区分不同特征的重要程度。可以使用信息论(information-theoretic methods)的一些方法来计算特征权重,比如信息增益(information gain)。本章对这些权重计算方法进行了具体介绍。此外,在k-NN实现上,本章对k这个超参数的选择进行了简要讨论,并强调,在k-NN的分类决策中,需要考虑到近邻实例和待分类实例之间的距离,使用这个距离来对分类投票(voting)进行加权,从而发展为距离加权的分类投票方法(distance-weighted class voting)——章节3.2.7对此方法进行了详细介绍。

基于一个现实的语言处理任务——德语名词的复数词法分析(plural formation of German nouns),本章对基于记忆的自然语言处理技术进行了详细说明,同时还介绍了本书作者和其团队开发的基于记忆学习的软件包——TIMBL,并说明了如何在语言处理任务中使用该软件包。相信通过使用该软件包,读者可以对基于记忆的语言处理技术有一个更直观的了解。

此外,本章对TIMBL软件包的内容、运行时间、空间的复杂度等进行了客观分析。因为就是简单的存储实例,该软件"训练(training)"效率很高,可以获得O(N)的时间复杂度——N是实例个数。但是存储的效率较低,其空间复杂度是 O(N*F)——F 是特征个数。不幸的是,"测试(testing)"过程的时间复杂度很高,复杂度是 O(N*M)——M 是测试数据实例个数——因为每个测试实例都需要和内存中存储的所有训练实例进行比较,从而确定近邻的实例。本章只是客观分析了时间、空间复杂度上的优缺点,并强调,在后面的章节中会提出高效率的估算(approximation)算法,从而可以更有成效地降低时间、空间复杂度。

第四章:在语素—语音学上的应用

基于之前的方法学介绍,本章重点介绍相关方法在具体任务上的应用,特别是怎么把一个新任务在基于记忆的语言处理框架下进行建模。本章重点讨论语素—语音学这一特定的语言处理任务。语素(morphemes)通常是由一个或者多个音素(phonemes)构成,且通常具有语义。在英语等西方语言的处理任务中,一个重要任务就是从文本/语音数据流中识别音素和语素——把机器学习应用到自然语言处理的最早的研究之一就是音素转换问题。在实际的应用场合可以碰到两种类型的词——已登录词和未登录词。比如,可以从标注好了语素/音素的训练数据中提取出一个已登录词列表及相关的语素/音素信息,不在此列表中的词为未登录词。语素—音素识别的重点就是这些未登录词,因为任何一个训练数据都是有限的,常会碰到未登录词的语素/音素识别问题。本章介绍怎样通过基于记忆的语言处理方法来解决语素/音素识别问题——包括英语的词—音素自动转换问题,以及荷兰语的语素自动识别问题(Dutch morphological analysis)。

本章详细介绍了作者团队开发的音素转换系统如何用于语音转换,并且解释了基于记忆的学习方法很适合处理词—音素自动转换问题的原因——拼写相似的英文词汇往往具有相似的发音,所以,对于一个新词,只需要在"记忆"中寻找其近邻的实例,从而就可以对其发音进行分类/预测。在荷兰语的语素自动识别方面,本章也介绍了详细的特征设置,比较了在基于记忆的语言处理框架下实现的 IB1 系统和 IGTree 系统——IGTree 系统是对传统 k-NN 算法的一种决策树(decision tree)快速估算法——的实验效果。基于这两个任务的实验效果,本章总结了 IGTree 这种决策树估算法的效果:能够有效降低内存需求,并且显著提高分类效率。

此外,本章揭示了基于记忆的语言处理方法的一个显著缺点,就是无法对结构依赖(structural dependencies)进行建模——比如两个不同位置的语素分类可能具有较强的结构依赖,通过对结构依赖进行建模可以形成更合理的全局分类结果,提高分类准确率。本书将在第七章讨论基于记忆的语言处理中结构依赖的建模问题。

第五章:在浅层句法分析上的应用

跟前一章类似,本章继续介绍相关方法在语言处理任务中的应用,通过另一个代表性的语言处理任务——浅层句法分析(shallow parsing)——来阐述如何把一个语言处理任务通过基于记忆的语言处理这个框架进行建模。跟前一章的语言处理任务相比,浅层句法分析是个更复杂的任务,因为这个任务同时需要处理词性标注(part-of-speech tagging,词性包括名词、动词等)、块切分(chunking,比如切分出名词短语、动词短语等)和关系提取(relation finding,比如确定名词短语和句子中的主动词之间的句法关系)这三个子任务。本章把浅层句法分析建模为层次化的分类问题,并使用基于记忆的语言处理技术来分别解决这些层次化分类问题。

本书作者强调,在词性标注这个子任务上,基于记忆的语言处理方法取得了比 Brill [1994]提出的基于转换的学习方法(transformation-based learning)——这是当时应用比较广泛的一种方法——更好的效果。但是结果稍差于 Ratnaparkhi [1996]提出的最大熵分类方法(maximum-entropy method)。最终,基于记忆的方法在标准数据集 WSJ 上取得了 96.4% 的准确率并且在 LOB 数据集上取得了 97% 的准确率。

词性标注之后,下一步就是进行块切分。块切分同样可以转化为一个分类问题,具体来说,可以通过 Ramshaw & Marcus [1995]提出的 BIO 模式把切分问题转换为分类问题,然后使用基于记忆的学习方法进行自动分类。因为词性标注和块切分这两个任务密切相关,可以把这两个任务的类别标记合并起来变成一个综合的分类任务。实验表明,虽然合并这两个任务会潜在地导致数据稀疏,但实际上取得了更好的总体分类效果。跟前面两个任务一样,最后一个任务关系提取也可以建模为一个分类问题,并使用基于记忆的学习方法进行自动分类。

第六章:抽象和泛化

通过前面五章的介绍,读者应该对基于记忆的自然语言处理有了方法学上的理解,并在具体的语言处理任务上有了实践认识。本章进一步探讨更本质、更核心的问题——为什么基于记忆的语言处理适合语言处理任务?也就是说,基于记忆的语言处理到底好在哪里?

作者对比分析了"懒惰(lazy)"的学习方法和"急切(eager)"的学习方法,讨论了这两大类方法的特点,并总结了它们在语言处理数据中各自的优缺点。急切学习方法的哲学根源是西方中世纪开始提出的"奥卡姆剃刀(Ockham's razor)"科学原则,其主要思想是,"一个科学理论应该去除所有不必要的元素(delete all elements in a theory that are not necessary)"——也就是说,一个科学理论应该在不影响效果的前提下越简单越好(用现代的观点解释就是最大熵),因为越简单的东西泛化能力越强。在机器学习领域,"奥卡姆剃刀"这种哲学思想在20世纪发展为一个具体的机器学习原则,就是"最小描述长度原则(minimal description length, MDL)"[Rissanen 1983],而最小描述长度原则指导了一系列的机器学习模型,包括用途广泛的决策树模型[Quinlan 1993]和规则自动生成算法RIPPER [Cohen 1995]等。

但是,作者强调,懒惰学习方法,包括基于记忆的学习,不符合最小描述长度原则,因为记忆模型的描述长度是所有使用的内存,也就是所有存储起来的训练实例。从一定程度上来说,基于记忆的学习方法甚至可以说是"最大化描述长度"的一种方法。作者强调,作为一种特定的懒惰学习方法,基于记忆的语言处理方法的主要优点是能够把所有的实例存储在记忆中,从而不需要承受抽象化(abstraction)带来的效果损失——主要是分类的准确度。本章介绍了一系列的实验,展示了实验的效果,从而佐证了作者的观点——懒惰的学习方法,特别是基于记忆的学习方法,在典型的语言处理任务上往往具有比急切的学习方法更好的分类效果。这里一个重要的原因是自然语言处理数据的特点——跟别的领域的任务相比,语言处理的训练数据(语料库)往往标注得很仔细,经过仔细检查,噪音已经很少,基本上不需要抽象化来提高泛化能力。从另一个角度来说,也许基于记忆的学习方法在语言处理数据上已经符合"最小描述长度原则",因为错误率很低的语言处理数据已经无法"压缩信息",也就是说,原来的"数据长度"就已经是最小描述长度。

此外,本章还详细介绍了FAMBL软件包。该软件包采用了一种特殊的抽象化方法用于改进基于记忆的学习系统。作者强调,这种特殊的抽象化方法(careful abstraction)跟传统的抽象化方法不同,在基于记忆的学习框架下能得到较好的效果,同时降低内存代价。

第七章:扩展阅读

本章重点介绍基于记忆学习技术的两种扩展:一种是基于搜索的参数优化算法,一种是在基于记忆学习的框架下解决序列标注任务(sequential tagging)上的"近视(near-sightedness)"问题。在这里我们主要介绍如何解决基于记忆学习的"近视"问题——也就是序列标注任务上的结构化预测问题。很多传统的机器学习模型都是无结构化的模型,也就是说,无法进行结构化预测，这包括本书介绍的基于记忆的学习方法，以及支持向量机(support vector machines, SVM)等。如果要对一个自然语言的数据序列——比如字符序列、词序列——进行处理，非结构化模型需要利用一个接一个的局部窗口(window)进行局部的分类，每个局部分类得到一个标记，最后多次的局部分类加在一起得到一个完整的标签序列。问题是，每次的局部分类都是不准确的，因为都没有考虑周围标签的结构化依赖信息。为了解决该问题，更现代的机器学习方法发展为结构化的分类模型，比如条件随机场模型(conditional random fields)。这些结构化分类模型能够"同时"决定整个序列的标签，从而充分考虑标签之间的结构化依赖信息。

为了在基于记忆学习的框架下部分实现"结构化分类"，部分解决决策的"近视"问题，本章提出了两种方法。方法一是分类器的叠加(stacking)。先使用一个基于记忆学习的初级分类器进行分类，得到一个初级的标注序列。之后，使用一个基于记忆学习的高级分类器，以原有数据序列信息以及初级标注序列信息作为新的输入，得到新的决策序列。因为有初级标注序列的信息，可以部分实现结构化依赖的建模。方法二是基于"类别多元组"的分类模型。这个方法主要是把原始的标记细化为标记的多元组，从而把结构化依赖的信息记录在"类别多元组"里面。作者强调，这两种方法都能有效解决基于记忆方法的"近视"问题，获得更好的分类准确度。实验表明，在相关的自然语言处理任务上，方法一能够降低11%的错误率，方法二能够降低13%的错误率，方法一和方法二还可以结合起来使用，在不同的自然语言处理任务上降低15%到34%的错误率。

3 本书的特色和不足

本书有两个主要的特点。第一个特点是介绍的方法简单实用。跟很多现有的语言处理技术相比，"基于记忆的学习"包括两个主要方面：一是学习方面，即简单存贮记忆；二是解决新问题，即简单搜寻最相似记忆并适当修正。本书有配套的软件发布，在介绍了核心方法之后，又详细介绍了作者团队开发的基于记忆学习软件包TIMBL，更进一步凸显了本书的实用性。

第二个特点是逻辑清楚、通俗易懂。跟书中介绍的基于记忆的学习方法的特点一样，本书写得很实在，有一说一，有二说二，章节安排层次清晰，内容深入浅出。对相关方法的描述，以及对相关原理的解释都很直观。

当然，本书也存在一些局限和缺点。首先，没有和更前沿的机器学习方法进行比较。本书的写作时间是2005年，那时很多新的机器学习模型和语言处理方法还没有提出来,所以本书基本上只和比较老的方法、系统进行了比较,比如规则系统、决策树系统等。和现有的前沿方法相比,这些模型的精确度要低不少。书中也提到了某些前沿的机器学习模型,比如条件随机场模型,但是没有进行详细的比较。这一点比较遗憾。此外,本书虽然提到了一些基于记忆学习方法的优点,比如泛化能力方面,但是缺乏理论方面的研究和论证。现有的机器学习方法往往能够通过理论分析对泛化能力进行定量分析(后面的延伸阅读会给出一些参考文献),相比之下,本书在理论分析方面有一定的局限。

其次,本书介绍的方法存在内存开销大,相似度计算时间复杂度高的缺点。虽然后面的章节提出了一些方法来改善这两个问题,但是本质上来说问题还是存在。特别是目前的语言处理、机器学习面临的是比以前更大规模的数据,会导致内存开销大、相似度计算时间复杂度高的缺点更为明显,原来提出的一些解决方案会遇到很大的挑战。

4 延伸阅读建议

本书主要探讨了基于记忆的语言处理技术,主要内容分为两部分:基于记忆的机器学习技术和该技术在语言处理任务上的应用。延伸阅读,可以立足于机器学习和语言处理的新进展这两方面。

(1) 语言处理方面的延伸阅读

在语言处理方面,可以阅读如下相关书籍:

Manning, C. D., & Schütze, H. *Foundations of Statistical Natural Language Processing.* Cambridge Massachusetts: MIT Press. 1999.(中译本:《统计自然语言处理基础》,苑春法等译,电子工业出版社,2005。)

俞士汶 计算语言学概论,商务印书馆,2003。

宗成庆 统计自然语言处理,清华大学出版社,2008(第二版2013年出版)。

此外,还可以关注语言处理领域的相关学术论文。本书的相当一部分内容是介绍语言处理任务中的参数估算,以及如何在基于记忆学习的框架下解决"近视"问题。所以,作为扩展阅读,可以参阅相关学术论文,特别是语言处理任务中的参数估算问题,以及结构化预测模型在语言处理领域的应用(如何使用结构化预测模型解决"近视"问题):

语言处理中的参数估算问题:

Jianfeng Gao, Galen Andrew, Mark Johnson, and Kristina Toutanova. A comparative study of parameter estimation methods for statistical natural language processing. In Annual Meeting of the Association of Computational Linguistics (ACL), 2007.

结构化感知器以及在自然语言处理的应用：

Michael Collins. Discriminative training methods for hidden Markov models: Theory and experiments with perceptron algorithms. In Empirical Methods in Natural Language Processing (EMNLP), 2002.

(2) 机器学习方面的延伸阅读

在机器学习方面，可以阅读如下相关书籍：

Christopher M. Bishop. *Pattern Recognition and Machine Learning.* Springer New York. 2006.

还可以关注机器学习方面重要会议和期刊的学术论文。本书讨论的基于记忆的学习方法有不小的局限性，主要问题之一就是难以解决结构化预测的问题——虽然本书的第七章讨论了一些改进的方法，但是这些方法还是不够系统，具有较大局限性。系统的解决办法就是采用结构化预测模型，最经典的结构化预测模型是条件随机场[Lafferty et al. 2001]，读者有兴趣的话可以阅读相关论文。此外，本书讨论了基于记忆学习方法的泛化能力的相关问题，但是缺乏系统的理论分析。在理论分析方面，读者可以阅读泛化能力分析相关的学术论文。在非结构化预测模型上，泛化能力分析的代表论文有Bousquet and Elisseeff [2002]。在结构化预测模型上，相关论文有Sun [2014]。此外，当前机器学习的一个热点是并行化机器学习，读者也可以阅读相关的文献，比如Niu et al. [2011]。

典型的结构化预测模型——条件随机场：

John Lafferty, Andrew McCallum, and Fernando Pereira. Conditional random fields: Probabilistic models for segmenting and labeling sequence data. In International Conference on Machine Learning (ICML), 2001.

传统机器学习模型的泛化能力分析：

O. Bousquet and A. Elisseeff. Stability and generalization. *Journal of Machine Learning Research.* 2002.

结构化预测模型的泛化能力分析，以及结构正则化——解决结构过拟合问题：

Xu Sun. Structure regularization for structured prediction. In Neural

Information Processing Systems (NIPS), 2014.

机器学习参数训练过程中的并行计算技术:

F. Niu, B. Recht, C. Re, and S. J. Wright. Hogwild: A lock-free approach to parallelizing stochastic gradient descent. In Neural Information Processing Systems (NIPS), 2011.

读者也可以通过相关领域的学术会议和期刊了解包括"基于记忆的语言处理"技术在内的通用语言处理技术、机器学习技术的最新进展。在语言处理／计算语言学方面,代表性的国际学术会议有 ACL(Annual Meeting of the Association for Computational Linguistics)、COLING (International Conference on Computational Linguistics)、EMNLP (Empirical Methods in Natural Language Processing);国内的相关学术会议有中文信息学会(CIPS)主办的"全国计算语言学会议(China National Conference on Computational Linguistics, CCL)";代表性的期刊有国际计算语言学学会主办的季刊 *Computational Linguistics*。机器学习方面的代表性国际学术会议有 ICML (International Conference on Machine Learning)、NIPS (Neural Information Processing Systems);代表性的期刊有 *JMLR* (*Journal of Machine Learning Research*)。

Preface

This book is a reflection of about twelve years of work on memory-based language processing. It reflects on the central topic from three perspectives. First, it describes the influences from linguistics, artificial intelligence, and psycholinguistics on the foundations of memory-based models of language processing. Second, it highlights applications of memory-based learning to processing tasks in phonology and morphology, and in shallow parsing. Third, it ventures into answering the question why memory-based learning fills a unique role in the larger field of machine learning of natural language – because it is the only algorithm that does not abstract away from its training examples. In addition, we provide tutorial information on the use of TIMBL, a software package for memory-based learning, and an associated suite of software tools for memory-based language processing.

For us, the direct inspiration for starting to experiment with extensions of the k-nearest neighbor classifier to language processing problems was the successful application of the approach by Stanfill and Waltz to grapheme-to-phoneme conversion in the eighties. During the past decade we have been fortunate to have expanded our work with a great team of fellow researchers and students on memory-based language processing in two locations: the ILK (Induction of Linguistic Knowledge) research group at Tilburg University, and CNTS (Center for Dutch Language and Speech) at the University of Antwerp. Our own first implementations of memory-based learning were soon superseded by well-coded software systems by Peter Berck, Jakub Zavrel, Bertjan Busser, and Ko van der Sloot. Ko invested the main effort in the development of the TIMBL software package since 1998. Ton Weijters cooperated with us in these early stages.

With other first-hour cooperators Gert Durieux and Steven Gillis we ventured into linking memory-based language processing to theoretical linguistics and to psycholinguistics, a line of research which was later continued with Masja Kempen, Evelyn Martens, and Emmanuel Keuleers in Antwerp. At the same time, Ph.D. students Jakub Zavrel, Jorn Veenstra,

Bertjan Busser, and Sabine Buchholz formed the initial team of the ILK research group in Tilburg with us.

Meanwhile we were fortunate to meet and be joined by colleagues elsewhere who worked on memory-based or related methods, and who were kind enough to reflect and relate our ideas to theirs on memory-based, instance-based, analogical, or lazy learning: notably David Powers, Jean-François Delannoy, David Aha, Dave Waltz, Claire Cardie, Hans van Halteren, Diane Litman, Koenraad De Smedt, Harald Baayen, Dominiek Sandra, Dietrich Wettschereck, Sandra Kübler, Yuval Krymolowski, Joakim Nivre, Hwee-Tou Ng, Daan Wissing, Christer Johansson, Anders Nøklestad, Joan Bresnan, David Eddington, and Royal Skousen, Deryle Lonsdale, and their colleagues.

In Tilburg and Antwerp we were subsequently joined by other cooperators, students, and postdoc researchers who helped shape and sharpen the algorithmic and methodological underpinnings of memory-based learning, as well as applied the method to many areas in natural language processing. We are very grateful to Véronique Hoste, Erik Tjong Kim Sang, Khalil Sima'an, Frank Scheelen, Guy De Pauw, Marc Swerts, Anne Kool, Stephan Raaijmakers, Piroska Lendvai, Emiel Krahmer, Martin Reynaert, Erwin Marsi, Laura Maruster, Olga van Herwijnen, Marie-Laure Reinberger, Fien de Meulder, Bart Decadt, Iris Hendrickx, Jacqueline Dake, Menno van Zaanen, Sander Canisius, Anja Höthker, An De Sitter, Jo Meyhi, Frederik Durant, and Kim Luyckx for their valuable contributions, and also to the other members of both research groups for being great colleagues even without doing memory-based learning.

This book has benefited greatly from the suggestions given by Maarten de Rijke, Valentin Jijkoun, Steven Bird, Helen Barton, Iris Hendrickx, Sander Canisius, Piroska Lendvai, and Anne-Marie van den Bosch.

Our research has been made, and continues to be made possible by our home universities, the University of Antwerp and Tilburg University, with major funding support from NWO, the Netherlands Organization for Scientific Research; FWO, the Flemish Organization for Scientific Research; IWT, the Institute for the Promotion of Innovation by Science and Technology in Flanders; the EU with its framework programmes; NTU, the Dutch Language Union, and the Netherlands Royal Academy of Arts and Sciences. We are grateful for their sustained trust and support.

Most trust and support came from our families. It is to this close circle of people that we dedicate this book.

Chapter 1

Memory-Based Learning in Natural Language Processing

This book presents a simple and efficient approach to solving natural language processing problems. The approach is based on the combination of two powerful techniques: the efficient storage of solved examples of the problem, and similarity-based reasoning on the basis of these stored examples to solve new ones.

Natural language processing (NLP) is concerned with the knowledge representation and problem solving algorithms involved in learning, producing, and understanding language. Language technology, or language engineering, uses the formalisms and theories developed within NLP in applications ranging from spelling error correction to machine translation and automatic extraction of knowledge from text.

Although the origins of NLP are both logical and statistical, as in other disciplines of artificial intelligence, historically the knowledge-based approach has dominated the field. This has resulted in an emphasis on logical semantics for meaning representation, on the development of grammar formalisms (especially lexicalist unification grammars), and on the design of associated parsing methods and lexical representation and organization methods. Well-known textbooks such as Gazdar and Mellish (1989) and Allen (1995) provide an overview of this 'rationalist' or 'deductive' approach.

The approach in this book is firmly rooted in the alternative empirical (inductive) approach. From the early 1990s onwards, empirical methods based on statistics derived from corpora have been adopted widely in the field. There were several reasons for this. Firstly, computer processing and storage capabilities had advanced to such an extent that statistical pattern recognition methods had become feasible on the large amounts of text and speech data gradually becoming available in electronic form. Secondly, there had been an increase of interest within NLP (prompted by application-oriented and competitive funding) for the development of methods that scale well and can be used in real applications without requiring a complete syntactic and semantic analysis of text. Finally, simple probabilistic methods had been enormously successful in speech technology and information retrieval, and were therefore being transferred to NLP as well. See Brill and Mooney (1998b) and Church and Mercer (1993) for overviews of this empirical revolution in NLP. The maturity of the approach is borne out by the publication of dedicated textbooks (Charniak, 1993; Manning & Schütze, 1999) and by its prominence in recent speech and language processing textbooks (Jurafsky & Martin, 2000) and handbooks (Mitkov, 2003; Dale et al., 2000).

Comparing these empirical methods to the knowledge-based approach, it is clear that they have a number of advantages. In general, probabilistic approaches have a greater coverage of syntactic constructions and vocabulary, they are more robust (they exhibit graceful degradation in deteriorating circumstances), they are reusable for different languages and domains, and development times for making applications and systems are shorter. On the other hand, knowledge-based methods make the incorporation of linguistic knowledge and sophistication possible, allowing them to be more precise sometimes. Probabilistic empirical methods are not without their own problems:

- The *sparse data problem*: often, not enough data is available to estimate the probability of events accurately. Some form of smoothing of the probability distributions is necessary.

- The *relevance problem*: it is often difficult to decide reliably on the importance or relevance of particular information sources for the solution of the NLP problem.

- The *interpretation problem*: most statistical techniques do not easily provide insight into *how* a trained statistical system solves a task. This is important for reasons of reusability, to integrate the acquired knowledge in existing systems, and to satisfy the end user's scientific curiosity about the linguistic tasks being performed by the trained system.

Symbolic machine learning methods in NLP offer interesting alternative solutions to these problems, borrowing some of the advantages of both probabilistic and knowledge-based methods. Some of these symbolic methods were created from within NLP (e.g., transformation-based error-driven learning, Brill, 1995), other techniques were imported from machine learning (Langley, 1996; Mitchell, 1997) into NLP; e.g., induction of decision trees and rules (Quinlan, 1993; Cohen, 1995), and inductive logic programming (Lavrac & Džeroski, 1994).

As a solution to the interpretation problem, rule induction methods, for example, present the learned knowledge in the form of condition-action rules of the type found in expert systems. This makes it easy for the domain expert to understand, evaluate, and modify them and to integrate them with hand-crafted rules. Inductive logic programming allows for the introduction of domain knowledge in the form of predicate logic expressions as background theories to the learning system, as a way to constrain the search for a model that covers the training examples. Rule induction and decision tree induction methods deal with sparse data and the information source relevance problem by relying on the minimal description length principle (Rissanen, 1983), a modern version of Ockham's razor, which states that smaller-sized models are to be preferred over larger models with the same generalization power. Rule learners and decision tree induction methods adopt this principle as a guideline in pruning decision trees, or finding small rule sets of simple rules that test only on those information sources that, according to reliability tests from statistics or information theory, are sufficient.

This book provides an in-depth study of memory-based language processing (MBLP), an approach to NLP based on a symbolic machine learning method called *memory-based learning* (MBL). Memory-based learning is based on the assumption that in learning a cognitive task from experience people do not extract rules or other abstract representations from their experience, but reuse their memory of that experience directly. Memory-based learning and problem solving incorporates two principles: learning is the simple storage of a representation of experiences in memory, and solving a new problem is achieved by reusing solutions from similar previously solved problems. This simple idea has appeared in many variations in work in artificial intelligence, psychology, statistical pattern recognition, and linguistics.

Over the last decade, we have applied this approach to many problems in NLP, and have come to the conclusion that the approach fits the properties of NLP tasks very well. The main reason is that for describing

NLP tasks mostly only a few clear generalizations can be found, with many conflicting sub-regularities and exceptions. Most learning methods are *eager*: they try to abstract theories from the data, and filter out exceptional, atypical, infrequent cases. We claim that in NLP, these cases constitute an important part of the required knowledge and should not be dismissed as noise. MBL is a *lazy* learning method which keeps all data available for processing and therefore also those cases from which more eager learning methods abstract. The eager versus lazy learning distinction will be described in more detail in chapter 6.

In the remainder of this chapter, we define NLP in a machine learning context as a set of classification tasks, introduce the terminology for MBLP we will use and expand throughout the book, and illustrate the approach by means of a well-known problem in NLP, *prepositional phrase attachment*. At the end of this chapter, we provide a roadmap for readers and explain why a relatively large part of the book is devoted to introducing software available with this book.

1.1 Natural language processing as classification

To enable the automatic acquisition of NLP knowledge from data we need machine-learning methods such as MBL, but in order to achieve any results, we must show first that NLP tasks can indeed be formulated in a machine-learning framework.

Tasks in NLP are complex mappings between representations, e.g., from text to speech, from spelling to parse tree, from parse tree to logical form, from source language to target language, etc., in which context plays an important, often crucial role. These mappings tend to be many-to-many and complex because they can typically only be described by means of conflicting regularities, sub-regularities, and exceptions. E.g., in a word processor's hyphenation module for a language such as Dutch (a simple and prosaic NLP problem), possible positions for hyphens have to be found in a spelling representation; the task is to find a mapping from a spelling representation to a syllable representation. Even this simple task is not trivial because the phonological regularities governing syllable structure are sometimes overruled by more specific constraints from morphology (the morphological structure of the word and the nature of its affixes). On top of that, there are constraints which are conventional, typographical, or which derive from the etymology of words. We find this interaction of constraints from multiple sources of information everywhere in NLP

1.1. NATURAL LANGUAGE PROCESSING AS CLASSIFICATION

and language technology; from phonological processing to pragmatic processing, and from spelling checking to machine translation. Linguistic engineering (hand-crafting) of a rule set and exception lists for this type of problem is time-consuming, costly, and does not necessarily lead to accurate and robust systems.

Machine learning (inductive learning from examples), by contrast, is fundamentally a *classification* paradigm. Given a description of an input in terms of feature-value pairs (a *feature vector*), a class label is produced. This class should normally be taken from a finite inventory of possibilities, known beforehand[1]. By providing a sufficient number of training examples (feature vectors with their correct class label), a machine learning algorithm can induce a *classifier*, which performs this mapping from feature vectors to class labels.

Suppose we want to learn how to predict the past tense of an English verb. This can be seen as a mapping from an input (infinitives of verbs) to an output (past tense forms of those verbs), e.g., work − worked, sing − sang. To redefine this mapping as a classification task, we have to transform the input into a fixed *feature vector*, for example by assigning a nominal feature to each part of the syllable structure (onset, nucleus, coda) of the infinitive. Each feature then represents a fixed part of the infinitive. The feature vectors associated with our two example words will become w,o,rk and s,i,ng, where commas indicate the boundary between features. We also need a finite set of possible output classes. For the past tense problem we might choose a class system consisting of -ed and for the irregular cases the particular vowel change involved (i-a, i-u, etc.). In order to train a system to learn this mapping, we need *examples*. Examples associate an *instance*, represented as a *feature vector*, with an output, represented by a *class* label. Feature values can be nominal (symbols), numeric, binary, or for some learning methods even complex and recursive. A few examples associated with our past tense task will now be represented like this:

---

[1]This distinguishes supervised classification-based learning from regression-based learning, where the class label is a real number.

Instance description			Class label
w	o	rk	-ed
s	i	ng	i → a
k	i	ll	-ed
sh	oo	t	oo → o

Different machine learning methods will use the information in the examples in different ways to construct a classifier from them. A memory-based classifier is trained by simply storing a number of instances with the correct class label in memory. A new instance of which the class is not yet known is classified by looking for examples in memory with a *similar* feature vector, and extrapolating a decision from their class. These examples are the *nearest neighbors* of the query instance. Chapter 3 provides a formal introduction to the algorithms and metrics that constitute our definition of MBLP, and that we will use in this book.

During the late 1990s the idea that all useful linguistic mappings — including complex (e.g., partially recursive) problems such as parsing— can be redefined as classification mappings and can thus be formulated in a ML context, has gained considerable support (Daelemans, 1995; Cardie, 1996; Ratnaparkhi, 1998; Roth, 1998). All linguistic problems can be described as classification mappings of two kinds: *disambiguation* and *segmentation* (Daelemans, 1995).

Disambiguation. Given a set of possible class labels and a representation of the relevant context in terms of feature values, determine the correct class label for this context. A prototypical instance of this situation is *part of speech tagging* (morphosyntactic disambiguation), a mapping in which a word that can have different morphosyntactic categories is assigned the contextually correct category. Other instances of this type of disambiguation include *grapheme-to-phoneme conversion, lexical selection in generation, morphological synthesis, word sense disambiguation, term translation, word stress and sentence accent assignment, accenting unaccented text, prepositional phrase attachment,* and *grammatical relation assignment*.

Segmentation. Given a target position in a representation and the surrounding context, determine whether a boundary is associated with this target, and if so, which one. A prototypical example here is *hyphenation* (or syllabification): given a position in a series of letters (or phonemes), a decision is made whether at that position a syllable boundary is present. Other examples include *morphological analysis*, *prosodic phrase boundary prediction*, and *constituent boundary detection*.

In such a perspective, complex NLP tasks such as *parsing* can be defined as a *cascade* of simpler classification tasks: segmentation tasks (finding constituent boundaries) and disambiguation tasks (deciding on the morphosyntactic category of words, the label of constituents, and resolving attachment ambiguities). We will return to a cascaded classification-based approach to shallow parsing in chapter 5. The classification-based approach reaches its limits, of course, at some point. Complex knowledge-based inference processes and semantic processes such as reasoning about tense and aspect would be formulated better in another way, although in principle a classification-based approach would be feasible.

An approach often necessary to arrive at the classification representation needed in this ML set-up for sequential tasks is the *windowing* method (as used in Sejnowski & Rosenberg, 1986 for text to speech). In this approach, a virtual window is shifted one item (e.g., word or phoneme) at a time over an input string (e.g., a sentence or a word). One item in the window, usually the middle item or the last item to enter the window, acts as the item in focus on which a disambiguation or segmentation decision is to be made, and the rest of the window acts as the context available for the decision. In this book we will discuss many examples of how to represent concrete NLP tasks within this classification-based learning approach.

1.2 A linguistic example

As a first, largely intuitive, illustration of how MBLP works, we develop its application to the well-known prepositional phrase disambiguation problem (*PP-attachment*)[2], where it has to be decided by a language understander which verb or noun is modified by a particular prepositional phrase. PP-attachment is a simple yet illustrative example of the pervasive problem of structural ambiguity, the resolution of which is essentially coupled to the proper understanding of the meaning of a sentence.

Arguably, memory traces of usage of earlier similar cases may help in the disambiguation. For example, in *eat a pizza with pineapple*, the

---

[2]This example is based on Zavrel et al. (1997).

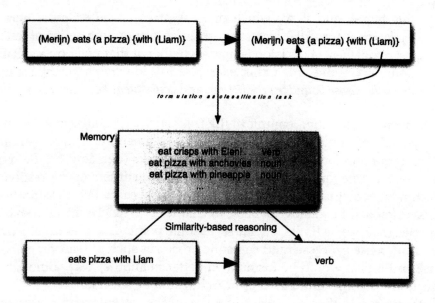

Figure 1.1: The general MBLP approach applied to PP-attachment. An NLP task is represented as a mapping between feature vectors and classes, learned through storage of examples in memory, and solved through similarity-based reasoning.

prepositional phrase with pineapple modifies pizza rather than eat because we have memory traces of similar expressions (e.g., eat a pizza with anchovies, eat a sandwich with cheese, ...) with the same interpretation. In eat a pizza with Eleni, other memory traces of similar sentence fragments such as eat crisps with Nicolas, and eat chocolate with Liam would favor a verb-modification interpretation. The feasibility of such a similarity-based approach depends crucially on a good operationalization of *similarity*, as well as sufficient relevant features to make a decision, and the availability of a sufficient number of examples.

Figure 1.2 sketches how the general MBLP approach introduced in the previous subsection would be applied to this problem. Given some NLP task (PP-attachment), the mapping is defined in terms of input features (heads of constituents, verbs, and prepositions) and output class (noun or verb attachment).

Several sources of information have been proposed to resolve prepositional phrase attachment ambiguity. Psycholinguistic theories have proposed disambiguation strategies which use syntactic information only,

1.2. A LINGUISTIC EXAMPLE

i.e., properties of the parse tree are used to choose between different attachment sites. Two principles based on syntactic information are minimal attachment (MA, construct the parse tree with the fewest nodes) and late closure (LC, attach as low as possible in the parse tree) (Frazier, 1979; Frazier & Clifton, 1998). Obviously, syntactic constraints only would never allow the correct resolution of all ambiguous examples discussed above. In those cases where syntactic constraints do not predict the attachment correctly, the meaning of the heads of the different phrases determines the correct attachment, and lexical information is essential in solving the task, as has been acknowledged in psycholinguistics as well (Boland & Boehm-Jernigan, 1998). Feasibility problems of hand-coding lexical semantic knowledge for a sufficiently large corpus have prompted the application of various corpus-based statistical and machine learning approaches to the PP-attachment problem (Hindle & Rooth, 1993; Brill & Resnik, 1994; Ratnaparkhi et al., 1994; Collins & Brooks, 1995; Franz, 1996; Zavrel et al., 1997; Stetina & Nagao, 1997; Abney et al., 1999). However, with the recent availability of lexical semantic resources such as FrameNet (Baker et al., 1998; Fillmore et al., 2003) and the Penn PropBank (Kingsbury et al., 2002), more sophisticated approaches are becoming available for encoding lexical semantic knowledge as features.

One straightforward way to apply the MBLP framework to this problem is to extract examples from sentences with a PP-attachment ambiguity such as the feature-vector and class representation shown in Table 1.1.

Llam ate his pizza with anchovies.
Nicolas eats crisps with Eleni.

verb	noun 1	preposition	noun 2	class
eat	pizza	with	anchovies	noun
eat	crisps	with	Eleni	verb

Table 1.1: Classification-based representation for the PP-attachment problem.

There is a benchmark data set[3] of PP-attachment examples, used first in Ratnaparkhi et al. (1994) which contains about 28,000 cases represented as feature vectors of the format displayed in Table 1.1. For each example, the verb, the head of the object NP, the preposition of the prepositional phrase,

---
[3]The data set has been made available by Adwait Ratnaparkhi from
ftp://ftp.cis.upenn.edu/pub/adwait/PPattachData

and the head noun of the NP of the prepositional phrase are recorded as feature values, and the class is either verb (attaching to the verb, or high attachment) or noun (attaching to the object noun, or low attachment). Applying an MBLP approach to this problem we learn from the data that the preposition is the most relevant of the four features, and that the other features are about equally relevant. On this particular data set, the correct attachment can be assigned for more than 80% of previously unseen examples by extrapolating from similar cases in memory. In chapter 3 we provide the details of our operationalization of this type of similarity-based reasoning. Here we briefly show by means of some of the test instances and their nearest-neighbor examples in memory how it can succeed, and how it can go astray.

The instance join board as director has a high attachment (as director attaches to the verb) which is correctly predicted by the classifier on the basis of examples such as name executive as director; elected him as senior director, etc. The instance changed face of computing is correctly classified as low attachment due to examples such as is producer of sauces; is co-author of books; is professor of economics and hundreds of equally similar examples in memory, effectively implementing a majority assignment for of as marking low attachment. We also see that sometimes analogy does not solve the problem, as with restrict advertising to places, which is incorrectly classified as a noun attachment case because of the single close noun attachment match regulate access to places.

For now, we hope to have exemplified how NLP problems can be formulated as classification tasks and how MBLP, working only on the basis of storing examples (for learning) and similarity-based reasoning (for processing) can achieve practical and interesting results.

1.3 Roadmap and software

This book tries to appeal to both computational linguists and machine learning practitioners, and especially to the growing group of colleagues and students who combine these two subfields of artificial intelligence in their research. For machine learning research, language processing provides a broad range of challenging problems and data sets. Features with thousands of values, tasks with thousands of classes and with millions of different examples, complex symbolic representations, complex cascades of classifiers; NLP has it all. For empirical computational linguistics research, (symbolic) machine learning approaches such as memory-based

1.3. ROADMAP AND SOFTWARE

learning offer additional tools and approaches on top of the statistical techniques now familiar to most computational linguists. We have tried to make the book accessible also to computational linguists with little previous exposure to symbolic machine learning, and to machine learning researchers new to NLP. We trust readers will find out quickly what to skip and what to dwell on given their previous knowledge.

After explaining the origins of memory-based language processing in linguistics, artificial intelligence, and psycholinguistics in chapter 2, chapter 3 dives into the details of our own specific approach to MBLP, discussing the algorithms and metrics used. For illustration and concreteness we provide a realistic problem, plural formation of German nouns, and a hands-on introduction to TIMBL, a software package available with this book. Distributed parts of the remainder of the book are devoted to explaining the use of TIMBL and a suite of associated programs and tools. For many of the NLP tasks described in later chapters, we also make available the data sets on which the results were achieved, or at least we provide clear pointers to how they can be acquired. The power and limitations of the memory-based approach can best be made clear by applying it to real problems. With TIMBL and the suite of associated software, we believe you will have an efficient and useful toolkit that you will be able to apply to real problems.

Chapters 4 and 5 discuss the application of MBLP to problems ranging from phonology to syntactic analysis, explaining the tasks, the results obtained with MBLP, and explaining the use of available software where relevant (e.g., IGTREE for phonology and MBT for part of speech tagging and chunking). The main goal of these two chapters is to provide worked-out examples on how to formulate typical NLP problems in an MBLP framework. Chapter 6 cuts to the core of why we think MBLP has the right bias for solving NLP tasks. We investigate the lazy–eager learning dimension and show empirically on a set of benchmark NLP data sets that language learning benefits from remembering all examples. Again, software is introduced that makes use of these insights by providing a careful, bottom-up approach to abstraction. Chapter 7, finally, introduces a number of current issues in MBLP (parameter optimization, alternative class representations, stacking of classifiers) that have a wider impact than just MBLP.

1.4 Further reading

Collections of papers on machine learning applied to NLP include Wermter et al. (1996); Daelemans et al. (1997b); Brill and Mooney (1998a); Cardie and Mooney (1999). Statistical and machine learning approaches are prominent in mainstream computational linguistics. Dedicated yearly conferences include EMNLP (Empirical Methods in Natural Language Processing, the yearly meeting of the SIGDAT special interest group of ACL) and CoNLL (Conference on Computational Natural Language Learning), the yearly meeting of the SIGNLL special interest group of ACL. Proceedings of these events are available on-line via the ACL Anthology[4]. SIGNLL also provides a growing number of data sets representing NLP tasks used in *shared tasks*, organized in alignment with CoNLL, which are intended to increase insight into which information sources and which machine learning methods work best for particular tasks.

---

[4] http://www.aclweb.org/anthology

Chapter 2

Inspirations from linguistics and artificial intelligence

Memory-Based Language Processing, MBLP, is based on the idea that learning and processing are two sides of the same coin. Learning is the storage of examples in memory, and processing is similarity-based reasoning with these stored examples. Although we have developed a specific operationalization of these ideas, they have been around for a long time. In this chapter we provide an overview of similar ideas in linguistics, psychology, and computer science, and end with a discussion of the crucial lesson learned from this literature, namely, that generalization from experience to new decisions is possible without the creation of abstract representations such as rules.

2.1 Inspirations from linguistics

While the rise of Chomskyan linguistics in the 1960s is considered a turning point in the development of linguistic theory, it is mostly before this time that we find explicit and sometimes adamant arguments for the use of memory and analogy that explain both the acquisition and the processing of linguistic knowledge in humans. We compress this into a brief review of thoughts and arguments voiced by the likes of Ferdinand de Saussure, Leonard Bloomfield, John Rupert Firth, Michael Halliday, Zellig Harris, and Royal Skousen, and we point to related ideas in psychology and cognitive linguistics.

De Saussure and Bloomfield. In his book *Cours de linguistique générale*, Ferdinand de Saussure (1916) put forward the theory that linguistics (the study of language) naturally divides into a linguistics of *la langue*, viz. language as a social medium, independent of the individual, and a linguistics of *la parole*, the individual's language. De Saussure argues that while *la parole* governs the way that individuals (learn to) generate and interpret speech, *la langue* represents the common, general knowledge of all individuals about the *signs* of the language, i.e., the common symbols of speech and writing used within a language community, and the relations existing between these signs (De Saussure, 1916).

Two relations exist between signs: (i) *syntagmatic* relations between signs at the same level; e.g., between letters in a word; (ii) *paradigmatic* (also referred to as *associative*) relations between signs at different levels, e.g., between letters and phonemes (De Saussure, 1916). Chomsky later pointed out that this dichotomy naturally corresponds to the processes of *segmentation* for determining the neighborhood relations of (sequences of) linguistic symbols at the same level, and *classification* for determining the correspondences between (sequences of) linguistic symbols at different levels (Piatelli–Palmarini, 1980). This, in turn, relates directly to the notions of segmentation (identification of boundaries) and disambiguation (identification) as discussed in the previous chapter.

In addition to the static aspects of language, De Saussure had elaborate ideas on the dynamics of language generation and analysis, centering for a major part around the concept of analogy, which he claimed has two opposing effects: it preserves and it alters. The effect of its preserving power is seen in the generation of new words and sentences, where these new strings contain sequences of symbols that display highly similar distributions of strings to other data of the same language seen previously and elsewhere. The mirror side of its preserving power is its power to alter, witnessed in the creation-by-analogy of completely new sequences of linguistic symbols that enter a language as neologisms. In De Saussure's words, "To analogy are due all normal non-phonetic modifications of the external side of words" (De Saussure, 1916, p.161). De Saussure goes on to refer rather explicitly to a capacity of a language user to access some sort of memory to generate utterances:

> Any creation must be preceded by an unconscious comparison of the material deposited in the storehouse of language, where productive forms are arranged according to their relations. (De Saussure, 1916, p. 165).

2.1. INSPIRATIONS FROM LINGUISTICS

While De Saussure stressed the role of analogy, Leonard Bloomfield advocated the role of the related dynamic concept of induction –roughly, the distilling of general rules or principles from examples– as the driving force behind discovering principles and regularities in language. Bloomfield at one point stated that "The only useful generalizations about language are inductive generalizations." (Bloomfield, 1933, p. 20). Throughout his work, Bloomfield proclaimed (in hindsight, behaviorist) ideas on the nature of the language faculty as emerging from learned responses to stimuli. He expresses the hypothesis that in generation, speakers use "analogy" and "habits of substitution" (Bloomfield, 1933, pp. 275ff) to generate sentences they themselves have never heard before. He makes an important distinction between "regular" non-exact, creative analogical generation from examples, and "irregular" memory retrieval of examples: "Any form which a speaker can utter only after he has heard it from other speakers, is irregular." (Bloomfield, 1933, p. 276).

Firth and Halliday. While De Saussure and Bloomfield stress the importance of the analogical process, John Rupert Firth stressed in his work that having real-world data is central to the development of any model of language; "A theory derives its usefulness and validity from the aggregate of experience to which it must continually refer" (Firth, 1952, p. 168, in Palmer, 1969). Firth is often quoted as a source of inspiration for the current interest in statistical models for natural language processing (Manning & Schütze, 1999), due to his work on purely data-driven language models, e.g., collocation models. How words occur in their context (e.g., in sentences, texts) can be explained most purely by "the company they usually keep" (Firth, 1952, p. 106ff, in Palmer, 1969). This approach can be interpreted from a memory-based viewpoint as easily as it can be taken as inspiration for probabilistic models; the role of a word in a sentence can be extrapolated from the collection of contexts in which it is remembered to occur.

With Firth as mentor and with the concept of groundedness in real-life data as a basis, Michael Halliday developed the theory of *systemic functional grammar* (Halliday, 1961), which views language essentially by its function in the real world, and its interaction with social context. Halliday acknowledges the breadth of this scope, but argues that it is essential, being the only way that language use can ever be understood or predicted fully. The main inspiration from Halliday's work that carries over to ours is the argument that it might be necessary to encode context using features far beyond the surface level of language: pragmatic features, rhetorical features, and any other feature in the social context that might be of importance in a particular time frame and text genre or type of discourse.

Harris and Skousen. American linguist Zellig Harris' *distributional methodology* can be seen as one proposal for a computational operationalization of memory-based language processing. Its methodology is grounded in data and statistics. As Harris put it, "With an apparatus of linguistic definitions, the work of linguistics is reducible (...) to establishing correlations. (...) And correlations between the occurrence of one form and that of other forms yield the whole of linguistic structure." (Harris, 1940, p. 704). His concept of substitution grammars (Harris, 1951; Harris, 1957) is an example of how this methodology can lead to a processing model of language: by analogy, sequences of forms receive the same syntactic label as other sequences that share the same or similar contexts.

An important step further in the operationalization is proposed by Royal Skousen who details a symbolic computational model of analogical processing (programmable, and indeed implemented by several researchers), and demonstrates its application to language (Skousen, 1989). Skousen's argument starts with the observation that all dominant linguistic theories have assumed rules to be the only means to describe aspects of language. Instead, Skousen argues for an inherently analogical approach to language and language learning, and introduces a concrete definition of analogy not based on rules. Like MBLP, Skousen's approach does not differentiate, as mainstream linguistics does, between regular language (i.e., language obeying rules), and irregular language (i.e., exceptions to the rules); rather, it treats language in an essentially unbiased way. This lack of bias fits language better, as Skousen argues: in language, the differences between regular data and exceptions can be quite graded.

Thus, to predict language behavior, and to model language learning, all that is needed is a large database of examples taken directly from language use, and a generically applicable method for analogical modeling that is inherently inductive but avoids the induction of rules (Skousen, 1989; Derwing & Skousen, 1989).

More recently, Skousen and colleagues published an overview of the analogical modeling approach, as well as links to other exemplar-based approaches to language (Skousen et al., 2002). In its introduction, Skousen explains what distinguishes his analogical modeling approach

2.1. INSPIRATIONS FROM LINGUISTICS

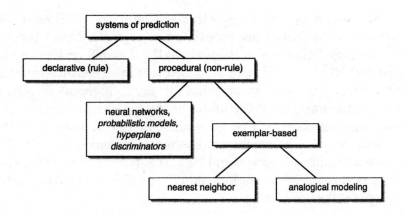

Figure 2.1: A categorical hierarchy of types of language prediction methods after Skousen et al. (2002).

from our nearest-neighbor approach. The differences are relatively small: they deal with the way the nearest neighbors are gathered from memory during classification, and how they are used together to produce classifications. Skousen gathers both approaches under the term "exemplar-based" when drawing the categorical hierarchy of "language prediction methods" displayed in Figure 2.1 (adapted from Skousen, 2002, p. 3). The hierarchy makes an opposition between exemplar-based methods and neural networks. To the latter group probabilistic methods can be added (e.g. Naïve Bayes classification, and maximum-entropy modeling), as well as other hyperplane discriminators (e.g., Winnow networks, kernel methods). The essential difference between these two groups of machine learning algorithms is the role of the exemplar (the instance) in analogical modeling and nearest-neighbor approaches; all other methods abstract away from the total set of examples.

Psychology and cognitive linguistics. Exemplar-based models have been proposed in psychology as well, more specifically in studies of human categorization, and often seem to produce a good fit of human behavior and errors (Smith & Medin, 1981; Nosofsky, 1986; Estes, 1994). These models assume that people represent categories by storing individual exemplars in memory rather than rules, prototypes, or probabilities. Categorization decisions are then based on the similarity of stimuli to these stored exemplars. More recent work seems to favor *hybrid* theories of categorization, i.e., the combination of rule-based or prototype-based and exemplar-based categorization (Johnstone & Shanks, 2001), or representational shifts from rule-based to exemplar-based processing as skills develop (Johansen & Palmeri, 2002). Overall, evidence for the psychological relevance of exemplar-based reasoning remains impressive. Recently, even the very assumption of *fixed*, permanent categories (however represented) has come under fire by theories favoring a *dynamic construal* approach in which concept formation is claimed to be based on past and recent experiences represented in memory, combined with current input (Smith & Samuelson, 1997). This type of context-dependent, memory-based category formation, akin to the ideas of Halliday referred to earlier, fits MBLP very well.

One recent approach to linguistics, *usage-based* models of language, represented by cognitive linguists such as Ronald Langacker, Joan Bybee, Arie Verhagen, William Croft and many others (see Barlow & Kemmer, 2000 for a collection of papers, and Croft & Cruse, 2003 for a textbook), bases itself at least in part on the psychological categorization literature and on some of the pre-Chomskyan linguistic approaches discussed earlier. It would lead too far to discuss the heterogeneous set of theories referred to by the label *usage-based*. Yet, some of the properties shared by them are reminiscent of the MBLP approach. Most importantly, the usage-based approach presupposes a bottom-up, maximalist, redundant approach in which patterns (schemas, generalizations) and instantiations are supposed to coexist, and the former are acquired from the latter. MBLP could be considered as a radical incarnation of this idea, in which *only* instantiations stored in memory are necessary. Other aspects of cognitive linguistics, such as the importance of frequency, experience-based language acquisition (Tomasello, 2003), and the interconnectedness of language processing with other cognitive processes fit MBLP as well.

In sum, the early twentieth-century linguistic literature on induction and analogy, culminating in Skousen's operational algorithm for analogical reasoning on the basis of exemplars, has been an important inspiration for the MBLP approach pursued in this book. As far as concrete operationalized algorithms are concerned, general (non-language-oriented) nearest-neighbor methods and their many descendents developed in artificial intelligence have played another driving inspirational role. It is to these methods that we turn next.

2.2 Inspirations from artificial intelligence

Nearest-neighbor classifier methods (most commonly named k-NN classifiers) were developed in statistical pattern recognition from the 1950s onwards (Fix & Hodges, 1951; Cover & Hart, 1967), and they are still actively being investigated in the research community. In these methods, examples (labeled with their class) are represented as points in an example space with as dimensions the numeric features used to describe the examples. A new example obtains its class by finding its position as a point in this space, and extrapolating its class from the k nearest examples in its neighborhood. Nearness is defined as the reverse of Euclidean distance. An early citation that nicely captures the intuitive attraction of the nearest-neighbor approach is the following:

> This "rule of nearest neighbor" has considerable elementary intuitive appeal and probably corresponds to practice in many situations. For example, it is possible that much medical diagnosis is influenced by the doctor's recollection of the subsequent history of an earlier patient whose symptoms resemble in some way those of the current patient. (Fix and Hodges, 1952, p. 43)

The k-NN literature has also generated many studies on methods for removing examples from memory either for efficiency (faster processing by removing unnecessary examples) or for accuracy (better predictions for unseen cases by removing badly predicting examples). See Dasarathy (1991) for a collection of fundamental papers on nearest-neighbor research.

However, the impact of k-NN methods on the development of systems for solving practical problems remained limited for a few decades after their development because of a number of problems. First of all, they were computationally expensive in storage and processing: learning reduces to storing all examples, and processing involves a costly comparison of an instance to all examples in memory. Interesting indexing and search pruning approaches were designed, such as k-d trees (Bentley & Friedman, 1979), but these only sped up the process for numeric features. A second problem with the approach was that the Euclidean distance metaphor for similarity breaks down with non-numeric and missing feature values. Further problems involved the sensitivity of the approach to feature noise and irrelevant features, and to the similarity metric used.

From the late 1980s onwards, the intuitive appeal of the nearest-neighbor approach promoted its adoption in artificial intelligence in many variations, using labels such as memory-based reasoning, case-based reasoning, exemplar-based learning, locally-weighted learning, and instance-based learning (Stanfill & Waltz, 1986; Cost & Salzberg, 1993; Riesbeck & Schank, 1989; Kolodner, 1993; Atkeson et al., 1997; Aamodt & Plaza, 1994; Aha et al., 1991). These methods modify or extend the k-NN algorithm in different ways, and aim to resolve the issues with the nearest-neighbor classifier listed before. As an example, the similarity metric for numeric features is extended in memory-based reasoning (Stanfill & Waltz, 1986) to nominal features, and in case-based reasoning (Riesbeck & Schank, 1989) even to complex recursive features such as graphs or trees.

The term *lazy learning* (as opposed to *eager learning*) has been proposed for this family of methods (Aha, 1997) because all these methods (i) delay processing of input until needed, (ii) process input by referring to stored data, and (iii) discard processed input afterwards. In contrast, eager learning methods abstract models (such as probability distributions or decision trees) from the examples, discard the examples, and process input by reference to the abstracted knowledge. Lazy learning methods have been applied successfully in robotics, control, vision, problem solving, reasoning, decision making, diagnosis, information retrieval, and data mining (see e.g. Kasif et al., 1998). The term lazy learning does not seem to have stuck. We have opted for memory-based language processing as a label to emphasize the role of the storage of all available data which we think is the main competitive advantage of this approach for NLP.

2.3 Memory-based language processing literature

Given the long tradition of analogical approaches in linguistics, even if not in the mainstream, their potential psychological relevance, and the success of memory-based methods in pattern recognition and AI applications, it is not surprising that the approach has also surfaced in natural language processing. Apart from the advantages inherent in all learning approaches, as discussed earlier (fast development, robustness, high coverage, etc.), advantages commonly associated with a memory-based approach to NLP include ease of learning (simply storing examples), ease of integrating multiple sources of information, and the use of similarity-based reasoning as a smoothing method for estimating low-frequency events. The latter property is especially important. In language processing tasks, unlike other typical AI tasks, low-frequency events that are *not* due to noise are perva-

2.3. MEMORY-BASED LANGUAGE PROCESSING LITERATURE

sive. Due to borrowing, historical change, and the complexity of language, most data sets representing NLP tasks contain many sub-regularities and exceptions. It is impossible for inductive algorithms to reliably distinguish noise from exceptions, so non-abstracting lazy memory-based learning algorithms should be at an advantage compared to eager learning methods such as decision tree learning or rule induction: 'forgetting exceptions is harmful'. We will return to this important characteristic of MBLP in later chapters.

Since the early 1990s, we find several studies using nearest-neighbor techniques for solving NLP disambiguation problems, framed as classification problems. A lot of this work will be referred to in the context of our own MBLP approach in later chapters. In the remainder of this one, we would like to point out similarities with two other popular approaches in NLP which are not further addressed in this book: example-based machine translation and data-oriented parsing.

Example-based machine translation (EBMT)

In seminal work, motivated from pedagogical principles in second language acquisition, Nagao (1984) proposed an approach to machine translation which is essentially memory-based. By storing a large set of (analyzed) sentences or sentence fragments in the source language with their associated translation in the target language as examples, a new source language sentence can be translated by finding examples in memory that are *similar* to it in terms of syntactic structure and word meaning, and extrapolating from the translations associated with these examples. Carl and Way (2003) and Jones (1996) provide overviews of different approaches within EBMT since its conception. In practice, because of the huge space of possible sentences to be translated, and the cost of collecting and searching large amounts of examples, EBMT systems are mostly hybrid, and contain rule-based as well as memory-based components. It is interesting to note that *translation memories*, arguably the most successful commercial approach to machine-aided translation today, are also based on the memory-based framework: large amounts of documents aligned with their translations are stored, and a possible translation for a new sentence is searched using approximate string matching techniques on sentences or sentence fragments.

Data-oriented parsing

Data-oriented parsing (DOP) is a statistical approach to syntactic parsing (Scha, 1992; Bod, 1998) that uses a corpus of parsed, i.e., syntactically or semantically analyzed utterances (a treebank), as a representation of a person's language experience, and analyzes new sentences searching for a recombination of subtrees that can be extracted from this treebank. The frequencies of these subtrees in the corpus are used to compute the probability of analyses. Such a method uses an annotated corpus as grammar, an approach formalized as stochastic tree substitution grammar (STSG). The advantage of STSG is that lexical information and idiomatic expressions (multi-word lexical items) can in principle play a role in finding and ranking analyses. Scha et al. (1999) provide an in-depth overview of the memory-based nature of the approach, tracing its motivation (as we do with MBLP) to pre-Chomskyan linguistics.

2.4 Conclusion

Models, both hand-crafted or learned, for an NLP task should be *general*: they should explain how to perform the task beyond a set of training examples or observations, and how to handle new, previously unseen cases successfully. In all areas of cognitive science, mechanisms of abstraction, especially hand-crafted rules, have become identified with this property of generalization by most researchers, and rote learning of examples, i.e. table lookup, is dismissed off-hand by many as missing this creative mechanism. In this chapter we have shown that this view has competing opinions. It makes sense to separate the dimensions of representation and learning. First of all, a rule-based approach does not have to be hand-crafted; rules can be learned. A probabilistic approach does not have to be based on training; probabilities can be assigned by hand.

But the main lesson learned in the literature described in this chapter is that generalization (going beyond the data) can also be achieved without formulating abstract representations such as rules: add an analogical reasoning mechanism to table lookup, and generalization becomes a property of the model. See Figure 2.2 for a visualization of the position of current cognitive science approaches in terms of generalization and abstraction. Abstract representations such as rules forget about the data itself, and only keep the abstraction. Such approaches are contrasted with table lookup, a method that obviously cannot generalize. However, by adding similarity-

2.4. CONCLUSION

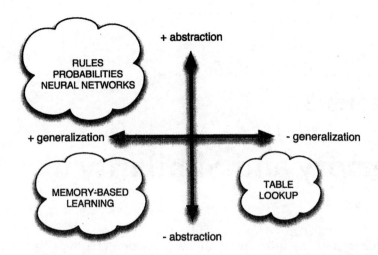

Figure 2.2: Generalization versus abstraction. Abstraction and generalization have been identified in most approaches with abtracting approaches pitted against table lookup. Extended with analogical reasoning, table lookup also generalizes.

based reasoning to table lookup, memory-based learning is capable of going beyond the data as well, and on top of that keeps all the data available. We will show that this is useful for NLP tasks: in such tasks, low-frequency or atypical examples are often not noise to be abstracted from in models, but on the contrary an essential part of the model.

In this chapter, we have shown that the idea of reducing learning to storage and processing to similarity-based extrapolation from stored memories is intuitively appealing for reasons of (psycho-)linguistic relevance as well as for reasons of computational accuracy and efficiency. The next chapter provides a detailed introduction to our own specific operationalization of these ideas in MBLP. As a linguistic example domain we have chosen the formation of the plural in German.

Chapter 3

Memory and Similarity

An MBLP system as introduced in the previous chapters has two components: a *learning component* which is memory-based, and a *performance component* which is similarity-based. The learning component is memory-based as it involves storing examples in memory (also called the *instance base* or *case base*) without abstraction, selection, or restructuring. In the performance component of an MBLP system the stored examples are used as a basis for mapping input to output; input instances are classified by assigning them an output label. During classification, a previously unseen test instance is presented to the system. The class of this instance is determined on the basis of an extrapolation from the most similar example(s) in memory. There are different ways in which this approach can be operationalized. The goal of this chapter is twofold: to provide a clear definition of the operationalizations we have found to work well for NLP tasks, and to provide an introduction to TiMBL, a software package implementing all algorithms and metrics discussed in this book[1]. The emphasis on hands-on use of software in a book such as this deserves some justification. Although our aims are mainly theoretical in showing that MBLP has the right bias for solving NLP tasks on the basis of argumentation and experiment, we believe that the strengths and limitations of any algorithm can only be understood in sufficient depth by experimenting with this specific algorithm.

To make our introduction more concrete, we will use *plural formation of German nouns* as a case study, based on Daelemans (2002). After an

---

[1] The TiMBL software and instructions on how to install it can be downloaded from the book's web site http://ilk.uvt.nl/mblp We assume a working version of TiMBL 5.0 or a later version has been installed on your system.

introduction to this problem, we provide a systematic introduction to the MBLP algorithms and their motivation, interleaved with tutorial material showing how to operate TIMBL on the plural formation data, and how to interpret its output. From this chapter onwards, TIMBL input and output is shown in outlined boxes, while practical discussions explaining the content of the boxes are printed around the boxes and are marked by a gray bar in the left margin. In this way, the theoretical and practical parts of the book are distinguished.

3.1 German plural formation

The diachrony of plural formation of German nouns has led to a notoriously difficult system, which is nevertheless acquired routinely by speakers of German. The complex interaction, from a synchronic point of view, of regularities, sub-regularities, and exceptions makes it a tough phenomenon to solve by means of handcrafted linguistic rules for use in morphological analysis and generation. Also from the point of view of cognitive modeling, the German plural is an interesting problem. Marcus et al. (Clahsen, 1999; Marcus et al., 1995) have argued that this task provides evidence for the *dual route* model for cognitive architectures. A dual route architecture supposes the existence of a cognitively real productive mental default rule, and an associative memory for irregular cases which blocks the application of the default rule. They argue that -s is the regular plural in German, as this is the suffix used in many conditions associated with regular inflection (e.g., neologisms, surnames, acronyms, etc.). This default rule is supposedly applied whenever memory lookup fails.

In this view, the case of German plurals provides an interesting new perspective to what is *regular*: the default rule (regular route) is less frequent than many of the 'irregular' associative memory cases. In a plural noun suffix type frequency ranking -s comes only in last place (after -(e)n, -e, - i.e., conversion; no suffix added, and -er, in that order). As most plural formation suffixes can be accompanied with a vowel change in the last stressed syllable (Umlaut), the variation we encounter in the data is considerable: Frau–Frauen (women), Tag–Tage (days), Sohn–Söhne (sons), Zimmer–Zimmer (rooms), Vater–Väter (fathers), Mann–Männer (men), Kind–Kinder (children), Auto–Autos (cars). "Now let the candidate for the asylum try to memorize those variations, and see how soon he will be elected." (Twain, 1880).

To prepare a data set for experimenting with MBLP on this problem,

Feature	Number of values	Kind	Auto	Vorlesung
Onset penultimate	78	-	-	l
Nucleus penultimate	27	-	aʊ	e
Coda penultimate	85	-	-	-
Onset last	84	k	t	z
Nucleus last	27	ɪ	o	ʊ
Coda last	79	nt	-	ŋ
Gender	10	N	N	F
Class	8	-er	-s	-en

Table 3.1: Some examples (instance as defined by the feature vector and associated class) for the German plural formation task.

we collected 25,753 German nouns from the German part of the CELEX-2 lexical database[2]. We removed from this data set cases without plurality marking, cases with Latin plural in -a, and a miscellaneous class of foreign plurals. From the remaining 25,168 cases, we extracted or computed for each word the plural suffix, the gender feature, and the syllable structure of the last two syllables of the word in terms of onsets, nuclei, and codas expressed using a phonetic segmental alphabet. Both the phonology and the gender of nouns are cited in grammars of German as important cues for the plural suffix. Table 3.1 gives an overview of the features, values, and output classes of the data. The gender feature has, apart from masculine (M), neuter (N), and feminine (F) also all possible combinations of two genders. The table also demonstrates three examples of the classification task to be learned. We will investigate whether this complex mapping can be learned with MBLP, and what we can learn about the problem.

3.2 Similarity metric

The similarity between a new instance X and all examples Y in memory is computed using a *similarity metric* (that actually measures distance) $\Delta(X, Y)$. Classification works by assigning the most frequent class within the k most similar example(s) as the class of a new test instance.

The most basic metric that works for instances with symbolic features

---

[2] Available from the Linguistic Data Consortium (http://www.ldc.upenn.edu/).

3.2. SIMILARITY METRIC

such as in the German plural data is the *overlap metric*[3] given in Equations 3.1 and 3.2; where $\Delta(X, Y)$ is the distance between instances X and Y, represented by n features, and δ is the distance per feature. The distance between two patterns is simply the sum of the differences between the features. In the case of symbolic feature values, the distance is 0 with an exact match, and 1 with a mismatch. The k-NN algorithm with this metric is called IB1 (Aha et al., 1991). Usually k is set to 1.

$$\Delta(X, Y) = \sum_{i=1}^{n} \delta(x_i, y_i) \qquad (3.1)$$

where:

$$\delta(x_i, y_i) = \begin{cases} \frac{x_i - y_i}{max_i - min_i} & \text{if numeric, otherwise} \\ 0 & \text{if } x_i = y_i \\ 1 & \text{if } x_i \neq y_i \end{cases} \qquad (3.2)$$

Our definition of this basic algorithm is slightly different from the IB1 algorithm originally proposed by Aha et al. (1991). The main difference is that in our version the value of k refers to k-nearest distances rather than k-nearest examples. Several examples in memory can be equally similar to a new instance. Instead of choosing one at random, all examples at the same distance are added to the nearest-neighbor set.

3.2.1 Information-theoretic feature weighting

The distance metric in Equation 3.2 simply counts the number of (mis)matching feature-values in two instances being compared. In the absence of information about feature relevance, this is a reasonable choice. Otherwise, we can add domain knowledge bias to weight or select different features (see e.g., Cardie, 1996 for an application of linguistic bias in a language processing task), or look for evidence in the training examples. We can compute statistics about the relevance of features by looking at which features are good predictors of the class labels. Information theory gives us a useful tool for measuring feature relevance in a way similar to how it is used as a tree splitting criterion for decision tree learning (Quinlan, 1986; Quinlan, 1993).

Information gain (IG) weighting looks at each feature in isolation, and estimates how much information it contributes to our knowledge of the

---

[3]This metric is also referred to as Hamming distance, Manhattan distance, city-block distance, or L1 metric.

correct class label. The information gain estimate of feature i is measured by computing the difference in uncertainty (i.e., entropy) between the situations without and with knowledge of the value of that feature (the formula is given in Equation 3.3), where C is the set of class labels, V_i is the set of values for feature i, and $H(C) = -\sum_{c \in C} P(c) \log_2 P(c)$ is the entropy of the class labels.

$$w_i = H(C) - \sum_{v \in V_i} P(v) \times H(C|v) \tag{3.3}$$

The probabilities are estimated from relative frequencies in the training set. For numeric features, an intermediate step needs to be taken to apply the symbol-based computation of IG. All real values of a numeric feature are temporarily discretized into a number of intervals. Instances are ranked on their real value, and then spread evenly over the intervals; each interval contains the same number of instances (this is necessary to avoid empty intervals in the case of skewed distributions of values). Instances in each of these intervals are then used in the IG computation as all having the same unordered, symbolic value per group. Note that this discretization is only temporary; it is not used in the computation of the distance metric.

The IG weight of a feature is a probability-weighted average of the informativeness of the different values of the feature. This makes the values with low frequency but high informativity invisible. Such values disappear in the average. At the same time, this also makes the IG weight robust to estimation problems in sparse data. Each parameter (weight) is estimated on the whole data set.

A well-known problem with IG is that it tends to overestimate the relevance of features with large numbers of values. Imagine that we would have the singular noun itself (e.g., Mann) as a feature in our German plural data set. Each value of this feature would be unique, and the feature will have a very high information gain, but it does not allow any generalization to new instances. To normalize information gain over features with high numbers of values, Quinlan (1993) has introduced a variant called *gain ratio* (GR) (Equation 3.4), which is information gain divided by $si(i)$ (split info), the entropy of the feature values (Equation 3.5).

$$w_i = \frac{H(C) - \sum_{v \in V_i} P(v) \times H(C|v)}{si(i)} \tag{3.4}$$

$$si(i) = H(V) = -\sum_{v \in V_i} P(v) \log_2 P(v) \tag{3.5}$$

The resulting gain ratio values can then be used as weights w_f in the weighted distance metric (Equation 3.6).

$$\Delta(X,Y) = \sum_{i=1}^{n} w_i\, \delta(x_i, y_i) \qquad (3.6)$$

The possibility of automatically determining the relevance of features implies that many different and possibly irrelevant features can be added to the feature set. This is a convenient methodology if domain knowledge does not constrain the choice enough beforehand, or if we wish to measure the importance of various information sources experimentally. However, because IG values are computed for each feature independently, this is not necessarily the best strategy. Sometimes more accuracy can be obtained by leaving features out than by keeping them with a low weight. Highly redundant features can also be challenging for IB1, because IG will overestimate their joint relevance. Imagine an informative feature which is duplicated. IB1 assigns the same weight to both copies, resulting in an overestimation of IG weight by a factor of two. This could lead to accuracy loss, because the doubled feature will dominate in the computation of the distance with Equation 3.6.

3.2.2 Alternative feature weighting methods

Unfortunately, as White and Liu (1994) have shown, the gain ratio measure still has an unwanted bias towards features with more values. The reason for this is that the gain ratio statistic is not corrected for the number of degrees of freedom of the contingency table of classes and values. White and Liu (1994) proposed a feature selection measure based on the χ^2 (chi-squared) statistic, as values of this statistic can be compared across conditions with different numbers of degrees of freedom.

The χ^2 statistic is computed from the same contingency table as the information gain measure by the following formula (Equation 3.7):

$$\chi^2 = \sum_i \sum_j \frac{(E_{ij} - O_{ij})^2}{E_{ij}} \qquad (3.7)$$

where O_{ij} is the observed number of cases with value v_i in class c_j, i.e., $O_{ij} = n_{ij}$, and E_{ij} is the expected number of cases which should be in cell (v_i, c_j) in the contingency table, if the null hypothesis (of no predictive association between feature and class) is true (Equation 3.8). Let $n_{\cdot j}$ denote the marginal for class j (i.e., the sum over column j of the table), $n_{i\cdot}$ the marginal for value i, and $n_{..}$ the total number of cases (i.e. the sum of all the cells of the contingency table).

$$E_{ij} = \frac{n_{.j} n_{i.}}{n_{..}} \qquad (3.8)$$

The χ^2 statistic is approximated well by the chi-squared distribution with $\nu = (m-1)(n-1)$ degrees of freedom, where m is the number of values and n is the number of classes. We can then either use the χ^2 values as feature weights in Equation 3.6, or we can explicitly correct for the degrees of freedom by using the *shared variance* measure (Equation 3.9):

$$SV_i = \frac{\chi_i^2}{N \times (min(|C|, |V_i|) - 1)} \qquad (3.9)$$

where $|C|$ and $|V_i|$ are the number of classes and the number of values of feature i, respectively, and N is the number of instances[4].

We turn now to the application of TIMBL to our German plural learning problem.

3.2.3 Getting started with TIMBL

TIMBL accepts data files in multiple formats. In this book we will presuppose a format where feature values and class are separated with spaces. E.g., the seven most frequent nouns according to the CELEX-2 lexical database, Jahr, Mensch, Zeit, Welt, Frage, Tag, and Land are stored as follows (in the DISC computer encoding of the International Phonetic Alphabet):

```
 -   -   -   j   a   r   N   e
 -   -   -   m   E   nS  M   en
 -   -   -   =   W   t   F   en
 -   -   -   v   E   lt  F   en
 fr  a   -   g   @   -   F   en
 -   -   -   t   a   k   M   e
 -   -   -   l   &   nt  N   Uer
```

---
[4]Note that with two classes, the shared variance weights of all features are simply divided by N, and will not differ relatively (except for the global division by N) from χ^2 weights.

3.2. SIMILARITY METRIC

Each noun is represented in terms of the syllable structure of the last two syllables, the gender, and the output class (the suffix, optionally preceded by the letter U if Umlaut occurs in the last stressed syllable).

The simplest TiMBL experiment is to divide the available data in two files (here `gplural.train` and `gplural.test`); train an MBLP system on the training set, and test how well it generalizes on unseen data in the test set. Machine learning methodology will be described in detail in section 3.5. For the experiment here, we randomly divided all examples into equal-sized training and test sets. On the command line (all example lines starting with '%' are commands issued on the command line; % represents the shell's prompt) the following command is issued:

```
% Timbl -f gplural.train -t gplural.test
```

This creates a file named `gplural.test.IB1.O.gr.k1.out`, which is identical to the input test file, except that an extra column is added with the class predicted by TiMBL. The name of the file provides information about the MBLP algorithms and metrics used in the experiment (the default values in this case). We will describe these and their meaning shortly.

Apart from the result file, information about the operation of the algorithm is also sent to the standard output. It is therefore useful to redirect the output to a file in order to make a log of this information.

```
% Timbl -f gplural.train -t gplural.test > gplural-exp1
```

We will now see what goes on in the output generated by TiMBL.

```
TiMBL 5.1.0 (release) (c) ILK 1998 - 2004.
Tilburg Memory Based Learner
Induction of Linguistic Knowledge Research Group
Tilburg University / University of Antwerp
Tue Dec 28 21:39:54 2004

Examine datafile 'gplural.train' gave the following results:
Number of Features: 7
InputFormat       : Columns
```

TiMBL has detected 7 features and the *columns* input format (space-separated features, class at the end).

```
Phase 1: Reading Datafile: gplural.train
Start:              0 @ Tue Dec 28 21:39:54 2004
Finished:       12584 @ Tue Dec 28 21:39:54 2004
Calculating Entropy        Tue Dec 28 21:39:54 2004
Lines of data    : 12584
DB Entropy       : 2.1229780
Number of Classes : 8

Feats   Vals    InfoGain        GainRatio
  1      67     0.14513201      0.031117798
  2      25     0.18009057      0.045907283
  3      77     0.15134434      0.060602898
  4      76     0.30203256      0.064526921
  5      23     0.68637243      0.20958072
  6      68     0.76659811      0.19727693
  7       9     0.69748351      0.45058817

Feature Permutation based on GainRatio/Values :
< 7, 5, 6, 2, 4, 3, 1 >
```

Phase 1 is the training data analysis phase. Time stamps for start and end of analysis are provided (wall clock time). Some preliminary analysis of the training data is done: number of training items, number of classes, entropy of the training data class distribution. For each feature, the number of values, and two measures of feature relevance are given (InfoGain, for information gain, and GainRatio). Shared variance or χ^2 are only displayed when selected. Finally, an ordering (permutation) of the features is given. This ordering is used for building an internal tree index to the instance base.

```
Phase 2: Learning from Datafile: gplural.train
Start:              0 @ Tue Dec 28 21:39:54 2004
Finished:       12584 @ Tue Dec 28 21:39:55 2004

Size of InstanceBase = 29509 Nodes, (590180 bytes),
52.50 % compression
```

3.2. SIMILARITY METRIC 35

Phase 2 is the learning phase; all training items are stored in an efficient way in memory for use during testing. Again timing information is provided, as well as information about the size of the data structure representing the stored examples and the amount of compression achieved.

```
Examine datafile 'gplural.test' gave the following results:
Number of Features: 7
InputFormat      : Columns

Starting to test, Testfile: gplural.test
Writing output in:          gplural.test.IB1.O.gr.k1.out
Algorithm       : IB1
Global metric   : Overlap
Deviant Feature Metrics:(none)
Weighting       : GainRatio
Feature 1              : 0.031117797754802
Feature 2              : 0.045907283257249
Feature 3              : 0.060602897678108
Feature 4              : 0.064526921164138
Feature 5              : 0.209580717876589
Feature 6              : 0.197276930994641
Feature 7              : 0.450588168767715

Tested:         1 @ Tue Dec 28 21:39:55 2004
Tested:         2 @ Tue Dec 28 21:39:55 2004
Tested:         3 @ Tue Dec 28 21:39:55 2004
Tested:         4 @ Tue Dec 28 21:39:55 2004
Tested:         5 @ Tue Dec 28 21:39:55 2004
Tested:         6 @ Tue Dec 28 21:39:55 2004
Tested:         7 @ Tue Dec 28 21:39:55 2004
Tested:         8 @ Tue Dec 28 21:39:55 2004
Tested:         9 @ Tue Dec 28 21:39:55 2004
Tested:        10 @ Tue Dec 28 21:39:55 2004
Tested:       100 @ Tue Dec 28 21:39:55 2004
Tested:      1000 @ Tue Dec 28 21:39:55 2004
Tested:     10000 @ Tue Dec 28 21:39:57 2004
Ready:      12584 @ Tue Dec 28 21:39:58 2004
Seconds taken: 3 (4194.67 p/s)
overall accuracy:       0.943659    (11875/12584), of which 6542 exact matches

There were 151 ties of which 85 (56.29%) were correctly resolved
```

In Phase 3, the trained classifier is applied to the test set. Because we have not specified which algorithm to use, the default settings are used (IB1 with the overlap similarity metric, information gain ratio feature weighting, and $k = 1$). Time stamps indicate the progress of the testing phase. Accuracy on the test set is logged, as well as the number of exact matches. Although there is no overlap in the words of the training and test set, the instances can nevertheless overlap (causing an exact match) when the last two syllables and the gender of different words are identical.

Finally, the number of correctly solved ties (two or more classes are equally frequent in the nearest-neighbor set) is given. In this experiment, the plural form of 94.4% of the new words was correctly predicted. Training and test set overlap in 6,542 instances (about half of the training data), and the algorithm had to break 151 ties, about half of which led to a correct classification.

The meaning of the output file names can be explained now: `gplural.test.IB1.O.gr.k1.out` means output file (`.out`) for `gplural.test` with algorithm IB1, similarity computed as *weighted overlap* (`.O`), relevance weights computed with *gain ratio* (`.gr`), and number of most similar memory items on which the output class was based equal to 1 (`.k1`).

3.2.4 Feature weighting in TIMBL

With option -w, we can investigate the effect of the different weighting methods described earlier on accuracy. The default is gain ratio (-w1). With -w0, -w2, -w3, and -w4 we can use no weighting, information gain, χ^2, and shared variance, respectively. Without weighting, accuracy drops to 91.8%; information gain does slightly better (on this data partition) than gain ratio with 94.6%, while χ^2, and shared variance perform again slightly better (94.8% both). From the point of view of understanding the effect of different information sources on solving the task, these feature relevance measures are very useful.

Figure 3.1 shows the distribution of three feature relevance weightings (scaled to $[0, 1]$) over the different features (syllable structure and gender). Clearly, the rhyme (i.e., nucleus and coda) of the last syllable is the most important information source for predicting the plural form. We see that gain ratio attaches more weight to the gender feature compared to information gain because the gender feature has fewer values than the syllable structure features. All metrics seem to agree that the information in the last but one syllable is not very relevant. TIMBL allows the experimenter to ignore features to test this hypothesis.

3.2. SIMILARITY METRIC

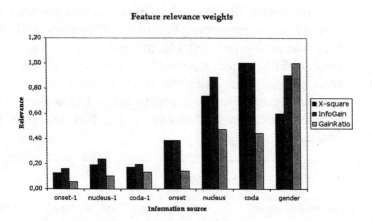

Figure 3.1: Three feature relevance weights (normalized to 1.0) for each of the seven features of the German plural data set.

```
% Timbl -f gplural.train -t gplural.test -mO:I1-3

Seconds taken: 1 (12584.00 p/s)
overall accuracy:        0.938334    (11808/12584), of which 11718 exact matches

There were 138 ties of which 77 (55.80%) were correctly resolved
```

-mO instructs TIMBL explicitly to use the (default) overlap metric. Separated by a colon, exceptions can be added, e.g., in this case to ignore (I) features one to three. With these settings, however, generalization accuracy diminishes considerably, and there are a lot more duplicate instances now. It is also possible to provide relevance weights (e.g., on the basis of linguistic intuition) directly to TIMBL via a weights file. Suppose a file gplural.weights is present with the following contents:

```
# Fea.  Weight
1       1
2       1
3       1
4       4
5       8
6       16
7       10
```

This represents the (linguistic) intuition that the final syllable is much more important than the last but one syllable, and that within the last syllable the nucleus and especially the coda (which together constitute the rhyme) are relevant, with gender being slightly more important than nucleus and clearly more important than coda. Using these hand-set weights yields the best result yet (on this data partition) of 95.0%.

```
% Timbl -f gplural.train -t gplural.test -w gplural.weights

Seconds taken: 3 (4194.67 p/s)
overall accuracy:         0.950095   (11956/12584), of which 6542 exact matches

There were 133 ties of which 69 (51.88%) were correctly resolved
```

3.2.5 Modified value difference metric

The choice of representation for instances in MBLP is the key factor determining the accuracy of the approach. The feature values and classes in NLP tasks are often represented by symbolic labels. The metrics that have been described so far, i.e., (weighted) overlap, are limited to either a match or a mismatch between feature values. This means that all values of a feature are seen as equally dissimilar to each other. However, in light of our German plural application we might want to use the information that the value neuter is more similar to masculine than to feminine in the gender feature, or that short vowels are more similar to each other than to long vowels. As with feature weights, domain knowledge can be used to create a feature system expressing these similarities, e.g., by splitting or collapsing features. But again, an automatic technique might be better in modeling these statistical relations.

For such a purpose a metric was defined by Stanfill and Waltz (1986) and further refined by Cost and Salzberg (1993). It is called the (modified) value difference metric (MVDM; equation 3.10), a method to determine the similarity of the values of a feature by looking at co-occurrence of values with target classes. For the distance between two values v_1, v_2 of a feature, we compute the difference of the conditional distribution of the classes $C_{1...n}$ for these values.

$$\delta(v_1, v_2) = \sum_{i=1}^{n} |P(C_i|v_1) - P(C_i|v_2)| \qquad (3.10)$$

3.2. SIMILARITY METRIC

MVDM differs considerably from overlap-based metrics in its composition of the nearest-neighbor sets. Overlap causes an abundance of ties in nearest-neighbor position. For example, if the nearest neighbor is at a distance of one mismatch from the test instance, then the nearest-neighbor set will contain the entire partition of the training set that contains *any* value for the mismatching feature. With the MVDM metric, however, the nearest-neighbor set will either contain patterns which have the value with the lowest $\delta(v_1, v_2)$ in the mismatching position, or MVDM will select a totally different nearest neighbor which has less exactly matching features, but a smaller distance in the mismatching features (Zavrel & Daelemans, 1997).

In sum, this means that the nearest-neighbor set is usually much smaller for MVDM at the same value of k. In some NLP tasks we have found it useful to experiment with values of k larger than one for MVDM, because this re-introduces some of the beneficial smoothing effects associated with large nearest-neighbor sets.

3.2.6 Value clustering in TIMBL

Applied to our data set, we see that MVDM does not help a lot, even with higher values of k.

```
% Timbl -f gplural.train -t gplural.test -mM -w0
overall accuracy:        0.922282    (11606/12584), of which 6542 exact matches
% Timbl -f gplural.train -t gplural.test -mM -w0 -k5
overall accuracy:        0.910521    (11450/12584), of which 6542 exact matches
```

Although the MVDM metric does not explicitly compute feature relevance, an implicit feature weighting effect is present. If features are very informative, their conditional class probabilities will on average be skewed towards a particular class. This implies that on average the $\delta(v_1, v_2)$ will be large. For uninformative features, on the other hand, the conditional class probabilities will approximate the class prior probabilities, so that on average the $\delta(v_1, v_2)$ will be very small. Nonetheless, adding a feature weighting metric to MVDM often improves generalization accuracy. In the following, MVDM is combined with χ^2 feature weighting.

```
% Timbl -f gplural.train -t gplural.test -mM -w3
overall accuracy:      0.941354  (11846/12584), of which 6542 exact matches
```

One cautionary note about MVDM is connected to data sparseness. In many practical applications we are confronted with a limited set of examples, with values occurring only a few times or once in the whole data set. If two such values occur with the same class, MVDM will regard them as identical, and if they occur with two different classes their distance will be maximal. In cases of such extreme behavior on the basis of low-frequency evidence, it may be safer to back off to the overlap metric, where only an exact value match yields zero distance. TIMBL offers this back-off from MVDM to overlap through a frequency threshold, set with the -L option that switches from MVDM to the overlap metric when one or both of a pair of matched values occur less often in the learning material than this threshold.

Although MVDM does not seem to improve accuracy on our task, we can show the effect of implicit value clustering in Figure 3.2, which displays the result of a hierarchical clustering of the conditional class probabilities associated with each value for the gender and nucleus of last syllable features. These probabilities can be written to a file in Timbl with -U filename.

```
% Timbl -f gplural.train -t gplural.test -mM -U gplural.matrices
% cat gplural.matrices

Targets : en, -, e, Ue, Uer, s, er, U.

feature # 1 Matrix:
-       0.098   0.327   0.274   0.063   0.015   0.183   0.033   0.008
g       0.110   0.362   0.360   0.032   0.023   0.073   0.031   0.010
b       0.179   0.238   0.352   0.073   0.011   0.101   0.030   0.015
d       0.115   0.143   0.522   0.069   0.009   0.104   0.037   0.000
r       0.166   0.120   0.619   0.038   0.007   0.038   0.013   0.000
St      0.184   0.045   0.760   0.006   0.000   0.006   0.000   0.000
br      0.276   0.126   0.457   0.047   0.024   0.031   0.008   0.031
...
```

3.2. SIMILARITY METRIC

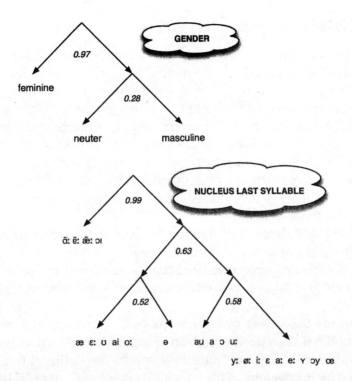

Figure 3.2: Hierarchical clustering of value-class conditional probabilities for gender feature and nucleus of last syllable feature. This clustering is implicitly used by MVDM to define similarity. The numbers indicate minimal distance between clusters scaled to $[0, 1]$.

The clustering for the gender (after mapping double gender assignments such as "Masculine or Feminine" to the most frequent gender) and nucleus of the last syllable shows that MVDM uses sensible phonological and lexical constructed knowledge implicitly. The lumping together of masculine and neuter gender and phonological categories such as front and back vowels do make sense and are sometimes used in morphological theories of German. The main advantage of MVDM is that these categories are automatically grouped in a task-dependent way, tuned to the task at hand, and arguably with more subtlety and more fine-grained than a representation in terms of phonological features and lexical classes.

3.2.7 Distance-weighted class voting

Until now, we have introduced two techniques that modify the effect of basic overlap similarity computation: the relevance of features, and the similarity between values of the same feature. A third parameter determining the outcome of MBLP systems is the number of nearest neighbors taken into account for extrapolation, which may depend on the density of the instance space. In dense spaces, even $k = 1$ can retrieve a considerable number of examples, in more sparse spaces, a higher value of k determines the robustness or smoothness of the extrapolation. Once a set of nearest neighbors is determined, there are different ways in which the output class can be decided.

The most straightforward method for letting the k-nearest neighbors vote on the class of a new case is the *majority voting* method, in which the vote of each neighbor receives equal weight, and the class with the highest number of votes is chosen (or in case of a tie, some tie resolution is performed).

We can see the voting process of the k-NN classifier as an attempt to make an optimal class decision, given an estimate of the conditional class probabilities in a local region of the data space. The radius of this region is determined by the distance of the k-furthest neighbor (Zavrel & Daelemans, 1997).

Sometimes when k is small and the data is sparse, or the class labels are noisy, the "local" estimate is unreliable. As it turns out in experimental work, using a larger value of k can often lead to higher accuracy. The reason for this is that in densely populated regions, the local estimates become more reliable with a larger k, because they are smoother. However, when the majority voting method is used, smoothing can easily become over-smoothing in sparser regions of the same data set. The radius of the k-NN region can get extended far beyond the local neighborhood of the query point, but the far neighbors will have the same influence in voting as the close neighbors. This can result in classification errors that could have easily been avoided if the measure of influence had somehow been correlated with the measure of similarity. To remedy this, distance-weighted voting can be used.

A voting rule in which the votes of different members of the nearest-neighbor set are weighted by a function of their distance to the query was first proposed by Dudani (1976). In this scheme, henceforth referred to as inverse-linear (IL) , a neighbor with smaller distance is weighted more heavily than one with a greater distance: the nearest neighbor gets

3.2. SIMILARITY METRIC

a weight of 1, the furthest neighbor a weight of 0 and the other weights are scaled linearly to the interval in between (Dudani, 1976), as defined in Equation 3.11:

$$w_j = \begin{cases} \frac{d_k - d_j}{d_k - d_1} & \text{if } d_k \neq d_1 \\ 1 & \text{if } d_k = d_1 \end{cases} \quad (3.11)$$

where d_j is the distance to the query of the j'th nearest neighbor, d_1 the distance of the nearest neighbor, and d_k of the furthest (k'th) neighbor.

Dudani, 1976, in his Equation 2.3, further proposed the *inverse distance weight* (henceforth ID). In equation 3.12 a small constant ϵ is added to the denominator to avoid division by zero (Wettschereck, 1994).

$$w_j = \frac{1}{d_j + \epsilon} \quad (3.12)$$

Another weighting function considered here is based on the work of Shepard (1987), who argues for a universal perceptual law which states that the relevance of a previous stimulus for the generalization to a new stimulus is an exponentially decreasing function of its distance in a psychological space (henceforth ED). This gives the weighted voting function of equation 3.13, where α and β are constants determining the slope and the power of the exponential decay function (henceforth we keep $\beta = 1$).

$$w_j = e^{-\alpha d_j^\beta} \quad (3.13)$$

Note that in equations 3.12 and 3.13 the weight of the nearest and furthest neighbors and the slope between them depend on their absolute distance to the query. This assumes that the relationship between absolute distance and the relevance gradient is fixed over different data sets. This assumption is generally false; even within the same data set, different feature weighting metrics can cause very different absolute distances.

Figure 3.3 visualizes a part of the curves of ID and ED, the latter with α set to 1.0 (default), 2.0, and 4.0. ID assigns very high votes (distance weights) to nearest neighbors at distances close to 0.0, while the ED variants have a normalized maximum vote of 1.0, and a less acute slope at distance 0.0. ID and ED with high α both give large prominence to exact matches; in the case of ID, the prominence is excessive. On the other hand, when the nearest neighbor in a particular classification with $k > 1$ is not an exact match, the differences in votes between nearest neighbors become smaller with larger distances, both with ID and with ED.

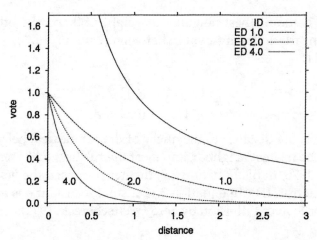

Figure 3.3: Visualization of the inverse distance weighting function (ID) and three variants of the exponential decay distance weighting function (ED) with $\alpha = 1.0, 2.0$, and 4.0.

Following Dudani's proposal, the benefits of weighted voting for k-NN have been discussed widely, e.g., in Bailey and Jain (1978); Morin and Raeside (1981); MacLeod et al. (1987), but mostly from an analytical perspective. With the popularity of memory-based learning applications, these issues have gained a more practical importance. In his thesis on k-NN classifiers, Wettschereck (1994) cites the inverse-linear equation of Dudani (1976), but proceeds to work with Equation 3.12. He tested this function on a large amount of data sets and found weak evidence for performance increase over majority voting. An empirical comparison of the discussed weighted voting methods in Zavrel (1997) has shown that weighted voting indeed often outperforms unweighted voting, and that Dudani's original method (Equation 3.11) mostly outperforms the other two methods.

3.2.8 Distance-weighted class voting in TiMBL

We have implemented in TiMBL the three types of *distance-weighted* voting functions described here. The default behavior is majority voting. Inverse distance, inverse linear, and exponential decay can be invoked with -d ID, -d IL, and -d EDa, respectively, in which the a in -d EDa is the parameter α.

3.3 Analyzing the output of MBLP

The different metrics discussed in section 3.2 provide a great deal of information about the problem being studied: feature weighting about the relevance of different information sources, and MVDM about the implicit clustering of feature values in a task-dependent way. In addition, individual classifications can reveal interesting details such as the actual nearest neighbors used; also, a fully classified test set offers the possibility to compute detailed statistics more informative than accuracy, the overall percentage of correctly classified test instances.

3.3.1 Displaying nearest neighbors in TIMBL

The TIMBL software includes a number of command-line switches that produce more detailed analyses of the results of classification. The +/-v (verbosity) option allows control of the output of different types of information. The most useful ones are the output of class distributions, distances, and nearest neighbors: +v db, +v di, and +v n, respectively, or combined: +v db+di+n.

```
% Timbl -f gplural.train -t gplural.test +v db+di+n
```

This generates, in the output file, for each test item a distribution of output classes, the distance to the nearest neighbor, and the examples on which the output was based (the nearest-neighbor set).

The following fragment of the output file illustrates this. The word Punch (tomfool) with plural Punchs is erroneously classified as having plural Punche. This is because there are four neighbors at distance 0.20, three of which (Part, Pack, and Patsch) have plural -e, while only one (Park) has the correct plural. Nearest-neighbor sets are useful in explaining the behavior of the classifier; the distance of the nearest neighbor can be used as a certainty factor. The distribution of output classes also allows back-off to a second solution and incorporation of the classifier in other modules which require statistical distributions instead of discrete solutions.

The distribution of classes in the nearest-neighbor set is *not* a statistical distribution, however, in the normal sense; rather, it is a local density distribution that cannot be compared directly to other local density distributions.

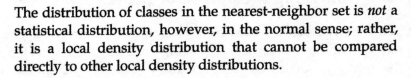

3.4 Implementation issues

Whereas the *k*-NN algorithm in its most naïve implementation is very fast in training ($O(N)$, where N is the number of training instances), as it consists only of storage and the computation of a few metrics, it is rather expensive in testing (namely $O(N * M)$, where M is the number of test instances): each test item has to be compared to all instances in memory to select the nearest neighbor(s). Storage requirements are in the order of $O(N * F)$, where F is the number of features. A fast approximation of MBLP with a much more attractive asymptotic complexity is described in the next chapter.

3.4.1 TIMBL trees

In the implementation of TIMBL, a tree-based memory structure is used in which examples are stored in a tree as paths from a root node to a leaf node, with arcs representing consecutive feature values (ordered according to a heuristic criterion), and with the leaf nodes representing a count of how many times which class occurs with the pattern of values corresponding with the path to that leaf node.

Because of the sharing of paths by instances, memory requirements are often considerably reduced, and processing time improved. E.g., for the German plural training instances, more than 50% compression of the training data is achieved. Information about additional methods to define similarity such as exemplar weighting, about the implementation of TIMBL, and a complete description of all parameters and their meaning can be found in the reference guide for TIMBL 5.1.0 (Daelemans et al., 2004b).

3.5 Methodology

For our German plural illustration, we have separated the available data into two equally sized data sets, one used for training and the other for testing. A more reliable methodology is to use *ten-fold cross-validation*, in which the available data are split into ten equal parts, and ten experiments are run each time taking out one part as test data and a concatenation of the remaining nine as training data. This way, all examples are used at least once as a test item, while keeping training and test data carefully separated, and the classifier is trained each time on 90% of the available training data rather than 50% as we have done until now in our examples. The average of the ten experiments provides a more reliable estimation of the true generalization accuracy on previously unseen data of our classifier (Weiss & Kulikowski, 1991). An additional advantage is that we learn about the variance in the generalization accuracy of the classifier by looking at the standard deviation in the results for the ten experiments. Taking this reasoning to the limit, an experimental methodology called *leave-one-out* can be employed which uses all available data except one example as training material, trains a classifier, and tests the classifier on the one held-out example, repeating this for all examples. In the leave-one-out experimental regime we use almost all available training data and will achieve, according to statistical theory, the most reliable generalization accuracy predictions.

3.5.1 Experimental methodology in TIMBL

In TIMBL, leave-one-out has been implemented efficiently by reusing data structures. Since leave-one-out does not assume a separate test set, in TIMBL's output file name the name of the absent test file is replaced with the string leave_one_out. We see that the overall accuracy is 95.1% with default settings, and that the time taken for doing the 25,168 experiments (gplural is the conjunction of gplural.train and gplural.test) is quite reasonable.

```
% Timbl -f gplural -t leave_one_out

Seconds taken: 60 (419.47 p/s)
overall accuracy:       0.950890    (23932/25168), of which 14759 exact matches

There were 335 ties of which 209 (62.39%) were correctly resolved
```

Another issue related to methodology is how to report generalization accuracy. Accuracy as we have used it until now is the number of times the trained system has predicted the same class (on the basis of the instance only) as the one present in the test set. This may be a bad measure of success in generalization for some data sets. Suppose, for example, that in German the -en plural would occur in 95% of the types, and the other suffixes in only 5% (showing skewed or ill-balanced class distributions, of which there are many examples in NLP), then an accuracy of 95% need not be impressive, as a simple rule always predicting -en would also achieve this accuracy without ever predicting a single non--en suffix correctly. This problem can be alleviated by reporting not only accuracy, but the complete *confusion matrix*, a table with the different classes both vertically and horizontally, associating the class predicted by the classifier with the real class of the test items given. From such a matrix not only accuracy can be derived, but also a number of additional metrics that have become popular in machine learning, information retrieval, and subsequently also in computational linguistics: *recall, precision,* and their harmonic mean *F-score; true positive*

3.5. METHODOLOGY

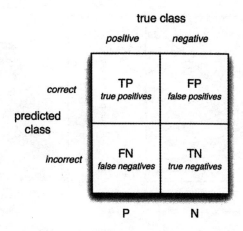

Figure 3.4: Class-specific confusion matrix containing the basic counts used in the advanced performance metrics.

rate, *false positive rate*, and their joint measure *area under the curve* (AUC) in receiver operator characteristics (ROC) space.

We describe these metrics in more detail. Figure 3.4 displays the general confusion matrix for one class C, splitting all classifications on a test set into four cells. The true positives (TP) cell contains a count of examples that have class C and are correctly predicted to have this class. The false positives (FP) cell contains a count of examples of a different class that the classifier incorrectly classified as C. The false negatives (FN) cell contains examples of class C for which the classifier predicted a different class. Finally, the true negatives (TN) cell contains examples with different classes that the classifier assigned a different class label than C. On the basis of these four numbers and the total number of positive examples $P = TP + FN$ and negative examples $N = FP + TN$, we can compute the following performance measures:

Precision $= \frac{TP}{TP+FP}$, or the proportional number of times the classifier has correctly made the decision that some instance has class C.

Recall or True Positive Rate (TPR) $= \frac{TP}{P}$, or the proportional number of times an example with class C in the test data has indeed been classified as class C by the classifier.

False Positive Rate (FPR) $= \frac{FP}{N}$, or the proportional number of times an example with a different class than C in the test data has been classified as class C by the classifier.

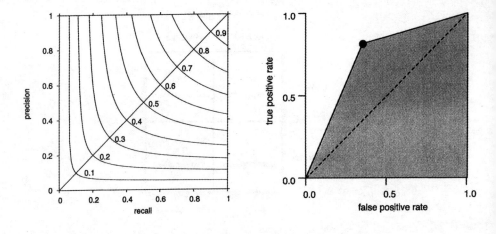

Figure 3.5: Precision–recall space with F-score isolines (left), and ROC space with an experimental outcome marked by the dot, and the outcome's AUC, the shaded surface between the dot and coordinates $(0,0)$, $(1,0)$, and $(1,1)$ (right).

F-score $= \frac{2 \times precision \times recall}{precision + recall}$, or the harmonic mean of precision and recall (Van Rijsbergen, 1979), is a commonly used metric to summarize precision and recall in one measure. The left part of Figure 3.5 shows F-score isolines in the two-dimensional space of recall (x-axis) and precision (y-axis). The curvature of the isolines is caused by the harmonic aspect of the formula (by contrast, the normal mean has straight isolines orthogonal to the $x = y$ diagonal), which penalizes large differences between precision and recall. The isolines could be likened to height isolines in a map, where the peak of the hill is at the upper right corner of the space.

AUC or *area under the curve* in the so-called ROC or *receiver operator characteristics* space (Egan, 1975; Swets et al., 2000), is the surface of the grey area in the right graph of Figure 3.5. The ROC space is defined by the two dimensions FPR (false positive rate, x-axis) and TPR (true positive rate, or recall, y-axis). The difference between F-score and AUC is that AUC does not make use of the statistically unreliable precision metric; rather, it takes into account all cells of the matrix in Figure 3.4 including the TN (true negative) cell (for a more detailed description and arguments for using ROC analysis, cf. Fawcett, 2004). Its "peak" is in the upper left corner, at a FPR of zero

3.5. METHODOLOGY

and a TPR of 1. Rather than using the harmonic mean, it is common to report on the AUC area under the classifier's TPR-FPR curve, where in the case of a discrete-output classifier such as TIMBL this can be taken to mean the two lines connecting the experiment's TPR and FPR to the $(0,0)$ coordinate and the $(1,1)$ coordinate, respectively; the AUC is then the grey area between these points and coordinate $(1,0)$.

Inspecting the full confusion matrix across classes may reveal additional dependencies and confusion ("leakage") among pairs or groups of classes. In general, looking at confusion matrices and at derived measures such as per-class precision, recall or TPR, FPR, F-score, and AUC, allows the experimenter to discover subtle effects of changing algorithmic parameters (since these affect some classes in different ways than others; some may improve recall, others may improve precision) that would not be visible when generalization performance was only expressed in terms of accuracy.

Another way of avoiding misinterpretations of accuracy scores is to compare the accuracy of the system to a *baseline* accuracy. One often used baseline is the frequency of the most frequent class in the data set (simulating a system which always predicts the majority class; 47% for our data), or a weighted product of the frequencies of the classes in the data (simulating a random class generator biased by the probability of the different classes; 33% for our data), or the result of applying a simple heuristic, e.g., a linguistically informed one.

3.5.2 Additional performance measures in TIMBL

TIMBL is able to produce class-specific counts on true positives (TP), false positives (FP), false negatives (FN), and true negatives (TN), as well as on the derived measures precision, recall or true positive rate (TPR), false positive rate (FPR), F-score, and AUC, the area under the curve in ROC space. To produce these class-specific performance measures, the command-line verbosity option +v cs (for class statistics) can be selected. TIMBL can also provide a weighted average of F-score and AUC over all classes with the verbosity option +v as, and it is able to produce a full confusion matrix with the verbosity option +v cm. They can be invoked together in the following command line:

```
% Timbl -f gplural -t leave_one_out +v cs+as+cm

Scores per Value Class:
class    TP    FP     TN    FN   precision  recall(TPR)  FPR        F-score   AUC
-      4261   137  20629   141   0.968849   0.967969     0.006597   0.968409  0.9807
e      4278   442  20080   368   0.906356   0.920792     0.021538   0.913517  0.9496
en    11713   207  13041   207   0.982634   0.982634     0.015625   0.982634  0.9835
s       658   165  24036   309   0.799514   0.680455     0.006818   0.735196  0.8368
er      266    32  24849    21   0.892617   0.926829     0.001286   0.909402  0.9628
Ue     1879   190  22968   131   0.908168   0.934826     0.008205   0.921304  0.9633
Uer     637    54  24427    50   0.921852   0.927220     0.002206   0.924528  0.9625
U       240     9  24910     9   0.963855   0.963855     0.000361   0.963855  0.9817

F-Score beta=1, microav: 0.950375
F-Score beta=1, macroav: 0.914856
AUC, microav:            0.968682
AUC, macroav:            0.952622
overall accuracy:        0.950890     (23932/25168), of which 14759 exact matches

Confusion Matrix:
             -      e     en      s     er     Ue    Uer      U
       -------------------------------------------------------
     - |  4261     45     26     46      5      8      4      7
     e |    32   4278     81     79     24    118     34      0
    en |    28    105  11713     27      0     42      3      2
     s |    61    148     72    658      3     16      9      0
    er |     1     17      1      0    266      0      2      0
    Ue |     4     90     24     11      0   1879      2      0
   Uer |     2     37      3      2      0      6    637      0
     U |     9      0      0      0      0      0      0    240
   -*- |     0      0      0      0      0      0      0      0

There were 335 ties of which 209 (62.39%) were correctly resolved
```

3.5. METHODOLOGY

Under "Scores per Value Class", TIMBL lists all per-class statistics in tabular format. From this table we see that suffix -en is very accurately predicted, while the infamous -s suffix is predicted with the lowest F-score (73.5%) and AUC (83.7%). A closer look at the row of suffix -s reveals that its precision (80.0%) is higher than its recall (68.0%). This indicates that TIMBL has predicted the -s suffix too conservatively; it was not predicted often enough at points where it should have been, but when it predicted it, it was fairly correct (80%).

Subsequently, TIMBL presents micro and macro averages of the F-score and AUC of all classes. The micro-average weighs all individual F-scores and AUCs proportionally to each class's frequency; the macro average is the unweighted average (and is usually the lower one, since low-frequency classes, which are typically predicted at lower generalization performance rates, receive equal weight in the macro average). The confusion matrix displays the predicted classes horizontally in the rows, and the real classes (the classes in the test set) vertically in the columns. In the experiment of which the confusion matrix is shown, TIMBL mistook the hard suffix -s most often for -e (79 times). There is a relatively high degree of leakage (mutual misclassification errors) between -s and -e: the most common misclassification of target class -e is -s (148 times), accounting for almost half the total precision error on suffix -s.

So far we have assumed that we choose either a fixed train-test split, or a cross-validation approach, with leave-one-out as cross-validation in the limit. A yet unspecified dimension is the size of the training set to begin with. With a small training set of, say, 200 words, the German plural formation task is likely to be harder to learn and apply to new unseen data, than with a training set consisting of the whole of CELEX-2. In many real-world situations the amount of annotated material is small or is slowly growing through time because it is being annotated by hand (or with the aid of machine learning tools that aid human annotation, cf. Thompson et al., 1999); in such situations it is valuable to not only have snapshot experimental scores, but also *learning curves*, i.e., series of experimental outcomes on systematically increased amounts of training

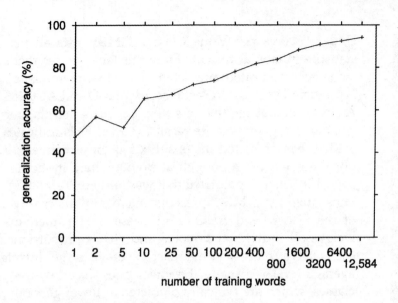

Figure 3.6: Learning curve of generalization accuracies on the German plural formation test set with increasing numbers of training words. The x axis has a logarithmic scale.

material up to the currently available total. To measure learning curves, it is common to take aside a fixed test set against which the increased model is systematically tested.

Figure 3.6 displays the learning curve on the German plural test set, with increasing amounts of instances in the training set (subsets were simply created by taking the n first instances off the full training set). The x axis has a logarithmic scale. The curve shows that at the maximal amount of training material currently available (12,584 words in the 50% split training set) the curve has not flattened; with larger amounts of training material better generalization accuracies on unseen data can be expected. On several occasions, almost log-linear curves have been reported on other NLP tasks (Banko & Brill, 2001; Van den Bosch & Buchholz, 2002), and the current curve is not far off from being log-linear either: it shows a fairly constant increase in generalization accuracy whenever the amount of training material is doubled, except for some understandable deviating behavior at the smallest numbers of training words. The curve suggests that if the training set is doubled in size about three or four times more, a 100% score on the test set appears to be in reach (save for some unsolvable ambiguities that would need more features, e.g., representing whether the word is a proper name). A triple doubling would mean an increase of the training set to over 100,000 words. An interesting question here would be if in the end most classifications would be based on exact matches; a point for future research.

A final note on methodology concerns significance tests. An experiment usually does not stand on its own; the experimenter is likely interested in comparing parameter settings, feature selections, or different learning algorithms on the same data. When some form of n-fold cross validation (CV) is performed, such as the common 10-fold CV, a pairwise comparative experiment yields two series of ten outcomes, each with an average and a standard deviation (see Dietterich, 1998, for an argumentation why a 2 × 5-fold CV setup is to be preferred over a simple 10-fold CV setup). Having paired sequences of cross-validation outcomes allows for the application of significance tests, to get a well-founded estimate of whether one variant is significantly better than the other.

Several significance tests may be applied to CV outcomes. Dietterich argues that paired t tests, on a meta-level, have a high recall in detecting statistical differences, but a somewhat low precision; paired t tests tend to make more Type-I errors (detecting a difference when it does not exist) than more conservative tests such as McNemar's test that do not need CV outcomes, or Dietterich's test for the 2 × 5-fold CV setup (Dietterich, 1998). Nevertheless, we briefly highlight the *paired t-test*, as arguably the simplest and most applicable test. The paired t-test (also the paired Student's t-test) assumes that the two sequences of outcomes are paired, i.e., that the two sequences are obtained on the exact same CV partitionings. It tests the null hypothesis that the population mean of the paired differences of the two sequences is zero. It produces a test outcome, the t value, which can be related to a threshold, the p value, that indicates the chances of the outcome of the test being false. Usually, $p < 0.05$ is regarded as a reliable sign that there is a significant difference between the two CV experiments, hence between the two variations (parameters, features, algorithms) tested. While it is common to apply t-tests in machine learning, it is much less common to go beyond pairwise statistical tests, e.g., to perform an analysis of variance (ANOVA) (but see for example Salzberg, 1997).

3.6 Conclusion

We mentioned in chapter 1 that the main attraction of symbolic machine learning methods in an empirical approach to NLP is that these may be able to alleviate some of the problems of statistical methods while keeping the advantages of robustness, coverage, reusability and accuracy compared to knowledge-based hand-crafting approaches. We have seen in this chapter that MBLP has advantages in all these areas: the similarity-based reasoning metrics described provide an implicit smoothing technique in sparse data. The feature-weighting methods provide an automatic approach to weighting the relevance of information sources. The study of nearest neighbors motivating a decision and the structure of the instance space (its density and homogeneity) provides tools for explanation and interpretation.

The elegant combination of the twin principles "learning is storage" and "reasoning is analogy" which define MBLP, has been shown to have a long history and strong defenders in linguistics, psychology, and artificial intelligence. We have combined and operationalized many of these ideas in a set of metrics and algorithms that includes automatic feature weighting, value clustering, and distance-weighted extrapolation, and provides a software tool, TIMBL, to experiment with these. We have found this combination of algorithms and metrics to work well in practice when trying to make efficient and accurate systems for solving NLP tasks, and we believe that the approach is necessarily the right one to choose in a machine learning approach to NLP, given the properties of problems in NLP. The first finding is further developed in chapters 4 and 5 where we show how MBLP can be applied to problems in phonology, morphology and syntax. The second hypothesis is explored further in chapter 6.

Chapter 4

Application to morpho-phonology

As argued in chapter 1, if a natural language processing task is formulated as either a disambiguation task or a segmentation task, it can be presented as a classification task to a memory-based learner, as well as to any other machine learning algorithm capable of learning from labeled examples. In this chapter as well as in the next we provide examples of how we formulate tasks in an MBLP framework. We start with one disambiguation and one segmentation task operating at the phonological and morphological levels, respectively.

A non-trivial portion of the complexity of natural languages is determined at the phonological and morphological levels, where phonemes and morphemes come together to form words. A language's phoneme inventory is based on many individual observations in which changing one particular speech sound of a spoken word into another changes the meaning of the word. A morpheme is usually identified as a string of phonemes carrying meaning on its own; a special class of morphemes, affixes, does not carry meaning on its own, but instead affixes have the ability to add or change some aspect of meaning when attached to a morpheme or string of morphemes.

One major problem of natural language processing in the phonological and morphological domains is that many existing sequences of phonemes and morphemes have highly ambiguous surface written forms, and especially in alphabetic writing systems where there is ambiguity in the relation between letters and phonemes. In English, for example, many letters, especially the vowels, can be pronounced in different ways; consider, for example, the pronunciation of ea in heart, bear, clear, and heard. Likewise, the word king can be seen in the words booking and looking, but neither word has a meaning that has anything to do with kings (normally speaking).

One seemingly simple solution to this is not to attempt to implement disambiguation systems, but to create big morphological and phonological lexicons (such as CELEX-2, Baayen et al., 1993) that cover a decent amount of a language's words, as well as predictable inflections of these words. This non-generalizing form of memory-based morphology and phonology, more aptly called rote-learned morphology and phonology, or lexical lookup, will guarantee a flawless lookup of the phonology and morphology of a language's most common words. Unfortunately, it is also guaranteed to fail in the case of unseen words and words for which syntactic knowledge (including contextual syntactic knowledge of the surrounding words) is needed to determine their appropriate pronunciation or morphological analysis. We defer the problem of syntactic processing to the next chapter; for now, the remaining issue is how to deal with unseen words.

Unseen words, which are likely to occur in unseen texts, often resemble seen words. As said, there is no guarantee that the unseen word will have the same pronunciation or the same morphological analysis as its nearest-neighbor known words. Yet, this is exactly what memory-based language processing assumes, and even if this may seem a strong assumption (especially when compared to hand-crafted rule-based methods that do not rely on lookup or similarity at all), the memory-based approach has offered state-of-the-art solutions in phonological and morphological processing. As examples of such solutions, in this chapter we demonstrate how memory-based language processing can handle English word phonemization and Dutch morphological analysis.

As we argue in this chapter, the key strength of the approach is to use representations of *parts of* words to perform memory-based reasoning on. The memory-based processing system will not find a reliable match with an unseen word but it is quite likely to find good matches between the unseen word's initial part and initial parts of known words; likewise for any other part. Phonological or morphological processing then becomes a memory-based generalized lookup process on word parts, and assembling the parts' outcomes to produce the full phonologically or morphologically processed output. Lexical lookup comes for free with this approach; if the system knows a word, it will look up the pronunciations or morphological analyses of its parts, and concatenate the parts' outcomes to reassemble the full pronunciation or morphological analysis.

4.1 Phonemization

We take phonemization to be the mapping of the orthography of unseen words to their pronunciation in some phonetic alphabet. Since new text will typically contain unknown words, it is vital in an application such as speech synthesis to be able to phonemize them at a reasonable level of accuracy (e.g., it is claimed that less than 5% error at the phoneme level is perceptually acceptable, Yvon, 1996).

In the traditional knowledge-engineering approach, several linguistic processes and knowledge sources are presupposed to be essential in achieving phonemization of unseen words. The classic MITALK text-to-speech system (Allen et al., 1987) exemplifies this multi-modular approach. The architecture of their phonemization module for unseen words includes explicitly implemented, linguistically motivated abstractions such as morphological analysis, rules for spelling changes at morphological boundaries, and phonetic rules. All these modules and their interaction are hand-crafted, and have to be redone for each new language. Also, lists of exceptions not covered by the developed rules have to be collected and listed manually.

The problems with such an approach include high knowledge engineering costs for the development of a single phonemization module, limited robustness and adaptability of modules once they are developed, and lack of reusability for other languages. As a consequence, data-driven learning methods using statistical or machine learning techniques have gained in popularity over the past two decades. Most of the earliest examples of machine learning applications to natural language processing are in fact phonemization (Stanfill & Waltz, 1986; Sejnowski & Rosenberg, 1987). Advantages of these systems include fast development times, high accuracy, robustness, and applicability to all languages for which data in the form of machine readable pronunciation dictionaries are available. We provide one example.

4.1.1 Memory-based word phonemization

The memory-based perspective to word phonemization is that the phonemization of a word, given its spelling, can be constructed by analogy from the phonemizations of words with similar spellings, of which the phonemization is known. Generally, words with similar spellings have similar phonemizations. The new word *veat* is pronounced [vit] because it resembles, among others, the known words *heat, feat, beat, eat, veal*, and seat in all of which the substrings ea are pronounced [i]. This type of memory-based reasoning has been acknowledged by many researchers as a useful approach in building phonemization modules (Damper, 1995; Yvon, 1996). It pays, however, to keep in memory the phonemizations of all known words, in order to prevent a word like great being pronounced as [grit] instead of [greɪt]. MBLP seems a viable approach here because (i) it keeps a lexicon of words with their associated phonemizations in memory, and (ii) it performs similarity-based reasoning to compute the phonemizations of words not in the lexicon.

4.1.2 TREETALK

The application of MBLP to letter-phoneme conversion was implemented in the TREETALK word phonemization system (Daelemans & Van den Bosch, 1996; Daelemans & Van den Bosch, 2001). Figure 4.1 lays out the architecture of TREETALK visually. Based on a lexicon of word-pronunciation pairs of a language L, a classifier is constructed that maps letters in their wordform context to phonemes. To construct the classifier, letters and phonemes of the word-pronunciation pairs in the lexicon are aligned, and subsequently these letter-phoneme conversion examples – as many examples per word as the number of letters – are compressed into a decision-tree structure (hence the name TREETALK) that approximates a normal k-NN classifier, but is much faster and less memory-intensive. We describe the alignment procedure (which is a necessary preprocessing step, involving expectation maximization, an unsupervised learning method), and then move on to pay special attention to the decision-tree compression module, IGTREE, which is a component of TIMBL.

Automatic alignment. The spelling and the phonetic transcription of a word often differ in length (e.g., rookie - [ruki]). Our phonemization approach demands, however, that the two representations be of equal length, so that each individual grapheme (one or more letters that are jointly mapped to one phoneme) can be mapped to a single phonetic symbol: r maps to [r], oo maps to [u], k maps to [k], and ie maps to [ie]. Our algorithm aligns the two representations by adding *null* phonemes in such a way that letters or strings of letters are consistently associated with the same phonetic symbols (e.g., rookie - [ru-ki-], where the hyphen depicts a phonemic null). Expectation maximization, EM (Dempster et al., 1977) is employed to create an optimized letter-phoneme probability matrix. The EM procedure is bootstrapped by first creating a probability matrix between letters and phonemes based on co-occurrence counts of letters

4.1. PHONEMIZATION

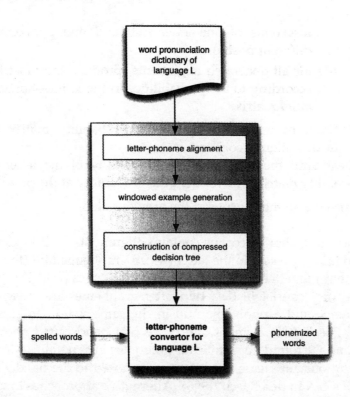

Figure 4.1: General architecture of the TREETALK word phonemization system.

and phonemes of word-phonemization pairs of equal length, i.e., where no alignment is needed. EM proceeds as follows:

1. Generate an aligned corpus: For each word-phonemization pair in the unaligned corpus,
   (a) when the phonemization contains fewer phonemes than the word contains letters, insert phonemic nulls at all possible combinations of positions in the phonemization to match the word's length;
   (b) when the phonemization contains more phonemes than the word contains letters, insert letter nulls at all possible combinations of positions in the word to match the phonemization's length;
   (c) when the number of letters equals the number of phonemes, generate the already aligned pair itself, but also all possible

insertions of one letter null and one phonemic null at different positions;

(d) for all generated alignments, produce the most likely one according to the probabilities in the letter-phoneme probability matrix;

2. Recompute the letter-phoneme probability matrix from the freshly aligned corpus;

Repeat until the total summed probability of the letter-phoneme probability matrix is lower than the probability at the previous step.

- Return the aligned corpus at the previous step.

The resulting aligned corpus contains alignments such as booking – [bukɪ-ŋ, and tax=i – [tæksɪ]. In the latter case, a graphemic null (the "=" sign) signifies that the letter x maps to the two phonemes [ks]. The EM alignments are very consistent (i.e., two-letter graphemes are always aligned to a phoneme and a phonemic null in the same way). In earlier work (Van den Bosch & Daelemans, 1993) we had handcrafted an alignment module, and we found no significant difference in generalization accuracy when using our automatic alignment as opposed to the handcrafted one (Daelemans & Van den Bosch, 1996). Alternative approaches to alignment are possible. For example, Luk and Damper (1996) combine stochastic grammar induction with Viterbi decoding (Viterbi, 1967) using a maximum likelihood criterion.

Windowed example generation. We deconstruct the word phonemization task into letter classification tasks: for each letter in a wordform, given its wordform context, we determine the phoneme label that this letter maps to. We define the phonemic mapping of a letter in context as its associated phoneme. Table 4.1 displays example instances and their phoneme label classifications generated on the basis of the sample word booking. For example, the first instance in Table 4.1, ___book, maps to class label [b].

In generating examples from words of which the wordform is shorter than the phonemization, such as the aforementioned taxi – [tæksɪ] which is aligned using a letter null as tax=i, the example with x in focus is mapped to the double phoneme class label "ks". These double phonemes take over the role of the letter null, which of course is not available in new words. In all other cases, such as the other letters of taxi and the examples generated from booking, the examples are labeled with the aligned phoneme or phonemic null. When a word's phoneme labels are predicted letter by

4.1. PHONEMIZATION

Instance number	Left context			Focus letter	Right context			Classification
1	_	_	_	b	o	o	k	b
2	_	_	b	o	o	k	i	-
3	_	b	o	o	k	i	n	u
4	b	o	o	k	i	n	g	k
5	o	o	k	i	n	g	_	ɪ
6	o	k	i	n	g	_	_	-
7	k	i	n	g	_	_	_	ŋ

Table 4.1: Examples generated for the word phonemization task, from the word-phonemization pair booking – [bukɪŋ], aligned as [b-ukɪ-ŋ].

letter, all non-null phonemes are concatenated to produce the complete word phonemization.

Construction of compressed decision tree. A normal memory-based system would store all letter–phoneme correspondences in memory. In each example the middle letter is surrounded by a fixed context of neighboring letters, that is wide enough to capture virtually all information for any letter-phoneme mapping to be unambiguous in the lexicon, given this context. However, the context that is sufficient for particular letter–phoneme mappings varies strongly. For example, in English, if a v is followed by an o it is most certainly phonemized as a [v]. For many other letters, particularly vowels, many more specific contexts are needed, often so specific that they only disambiguate between a few or just two wordforms; consider, for example, the phonemization of the i in reside versus resident, and of ea in hear, heard, bear, and beard.

It might be highly efficient to store examples in memory with just enough context to safely disambiguate the phonemization of the focus letter. Since no more context would be needed in the classification, it could also lead to a potentially large speedup when the classifier was told to return the assumedly correct phoneme, rather than spend time finding the exact k-nearest neighbors that match best over all features, including those outside the minimally sufficient context.

Such a classifier can be realized in a *decision tree*, for which machine learning offers standard induction algorithms (Breiman et al., 1984; Quinlan, 1986; Quinlan, 1993). A decision tree compresses a set of examples in such a way that only the most decisive information sources in examples (i.e., those features leading to a classification in the smallest number of steps) are retained in the tree. A typical decision tree, such as the trees produced by the C4.5 top-down induction of decision trees algorithm (Quinlan, 1993), consists of nodes and arcs; nodes represent subsets of the training set of examples, and store the most frequent classification in that subset. From the root node, arcs fan out representing the values of one feature on which the decision tree tests. Typically one arc tests the presence of one value at the tested feature. Each arc leads to one node, which represents the subset of examples that share the feature value tested on the arc that lead to the node. Nodes can either fan out further, or can lead to a leaf; typically, a leaf signals that the examples in the subset at that node all have the same class, i.e., are homogeneous with respect to their class labeling.

A typical top-down decision-tree induction algorithm attempts to construct trees in which the number of nodes is kept to a minimum. The most commonly used heuristic that is the key to this minimization is to test on the most informative, or class-discriminative features first. The classic ID3 algorithm (Quinlan, 1986) and its successor C4.5 (Quinlan, 1993) use information gain or gain ratio (see also subsection 3.2.1) to estimate the most informative features. Thus, the key difference with the IG or GR-weighted k-NN classifier is that these decision trees do not use IG and GR as weights (in a similarity function), but as rank numbers that determine the order in which features are tested in the tree. A decision tree variant that we introduce here, and which is included in the TIMBL package, is IGTREE (Daelemans et al., 1997b), a decision-tree approximation of IB1 that induces a decision tree using IG, GR or any other feature weight as feature ranking metric. IGTREE operates as follows.

As with standard IB1, we start by computing feature weights on the basis of a training set of labeled examples, as exemplified in chapter 3. Figure 4.2 displays example IG, GR and χ^2 feature weights, normalized to 1.0, of letters in their wordform context, represented by a window of three neighbor letters to the left and three neighbor letters to the right. The data in Figure 4.2 are computed on English letter-phoneme examples; we describe this task in more detail in the remainder of this section. Clearly, the focus letter is by far the most relevant feature according to all three functions. The explanation is straightforward; to phonemize a letter, it is more important to know which letter is to be phonemized, than to know the identity of its left or right neighboring letters. The focus letter reduces the subset of classes between which the classifier needs to decide, if it needs to decide at all; most letters map to just one or two possible phonemes, others

4.1. PHONEMIZATION

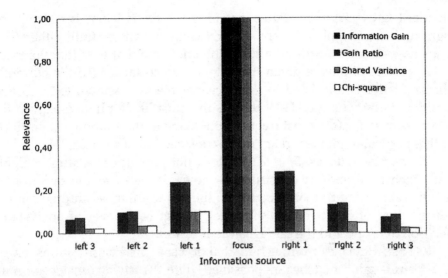

Figure 4.2: Information Gain, Gain Ratio and χ^2 feature weights (normalized to 1.0) for letter-phoneme classification information sources: the focus letter to be phonemized, surrounded by three left and right neighbor letters. The numbers are based on English letter-phoneme data.

to a handful. Context is of secondary importance, needed to disambiguate among alternatives.

The IG and GR feature weight of the focus letter is almost as big as the sum of all other IG or GR feature weights; with χ^2, the weight of the focus letter is more than twice as big as the sum of the rest. In other words, when a memory example mismatches on the focus letter, it is very unlikely that this example would be the nearest neighbor of a test instance, given that there will always be better-matching examples that do match at least on the focus letter. This justifies the use of an efficiency speedup in which the nearest neighbors (with $k = 1$) are sought only among the memory examples that have the same focus letter. Having to search among this smaller subset of possible nearest-neighbor candidates reduces the search for the nearest neighbor considerably. The efficient zooming-in on partitions of the training set can be repeated in the order of relevance of features. This, in essence, constitutes the IGTREE decision-tree induction algorithm.

Obviously, the weights of the subsequent context features in the example displayed in Figure 4.2 are not larger than the sum of the remaining features – this is where the IGTREE algorithm starts to diverge from the standard *k*-NN classifier. After matching the focus letter, it tests on the right and left neighbor letters, at increasing distances, until either (i) it finds an end node (*leaf*) that returns the phoneme that is at that point the only possible phoneme given the letters seen so far, or (ii) the currently checked letter is not listed in the decision tree as a known letter in that position, upon which IGTREE returns the most likely phoneme given the letters seen so far (the most frequent phoneme in the training set partition at that particular zoom-in point in the construction of the tree).

In contrast with C4.5, IGTREE does not recompute feature weights with each new node it constructs – rather, it sticks to the ordering of features according to their weights on the full training set throughout the construction of the tree. This implies that at each level of an IGTREE decision tree, one and the same feature is tested.

IGTREE deviates from IB1 in that it bases its classifications on as few feature-value matches as possible, while IB1 always matches on all feature values. This makes IGTREE an approximation of IB1; IGTREE will usually classify instances in less time than IB1 would, and may produce different classifications. Note that IGTREE and IB1 do share a common internal structure; as mentioned in section 3.4, IB1 compresses its training examples in the same decision tree structure, but uses this to search, with backtracking, for the *k*-nearest neighbors.

As an example, Figure 4.3 shows part of an IGTREE built for English phonemization. The path in bold represents the phonemization of a in the word **behave**. Ultimately, the tree checks whether the second letter to the left is an **e**, to disambiguate the context **ehave** from the different pronunciation of the **a** in the similar wordform **have**.

The result of this compression using IGTREE is that all transcriptions in the lexicon can be retrieved from this data structure very quickly, and that the lexicon is compressed into a tree (actually, a *trie*, Knuth, 1973) that takes up a fraction of the size of the original lexicon. Leaves near the top of the tree represent letter-phoneme mappings with limited context (reminiscent of the example rewrite rules given by Chomsky & Halle, 1968 in their classic work on English word pronunciation), while leaves further down the tree represent more specific, and eventually example-specific contexts for phonemization disambiguation. The compression rate depends on the size and coverage of the lexicon used for training, but certainly also on the language. Alphabetic writing systems with more or less one-to-one letter-phoneme mappings (*shallow orthographies*) tend to produce higher levels of compression than writing systems with complex mappings (*deep orthographies*), such as English (Van den Bosch et al., 1995).

4.1. PHONEMIZATION

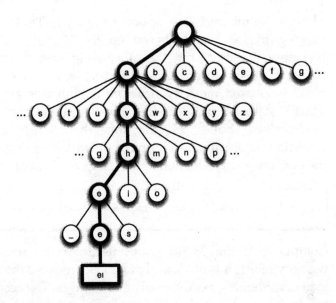

Figure 4.3: Retrieval of the pronunciation of a, of the word behave, in a partial depiction of the IGTREE for English phonemization. The path represents the minimally disambiguating context ehave when following the context letters in information-gain order.

4.1.3 IGTREE in TIMBL

The IGTREE decision tree induction algorithm is a component of the TIMBL package. IGTREE's feature ranking is set by the selected feature weighting parameter. This is the only parameter that influences IGTREE. It is known to compress data sets considerably, and speed up classification, usually at the cost of some generalization accuracy. Notable loss of accuracy as compared to standard IB1 is often reported in cases in which a task representation contains several important features with similar weights.

On the German plural data set used earlier in chapter 3, IGTREE runs as follows:

```
% Timbl -f gplural.train -t gplural.test -a1
```

The -a1 command line option triggers IGTREE. Feature ranking in the tree is based on the default GR feature weighting function, but the other weighting functions can also apply (-w). After all data has been read in memory and a full tree-based memory structure is built (see section 3.4), TIMBL starts pruning the tree down to the necessary paths as described above. Having done that, it reports on the compression obtained, in terms of the numbers of nodes and the percentage of reduction in the usage of bytes of memory:

```
Size of InstanceBase = 1311 Nodes, (26220 bytes), 97.87 % compression
```

Compared to the 28,515 nodes used for the tree-indexed memory of IB1, it is obvious that IGTREE reduces the memory load considerably, needing only 1,311 nodes. The accuracy on test instances shows that IGTREE is doing a little better than IB1 with GR feature weighting (11,849 out of 12,584 classified correctly, 94.16%):

```
Seconds taken: 2 (6292.00 p/s)
11872/12584 (0.943420)
```

This shows that IGTREE does not always decrease accuracy.

Classifier. The IGTREE decision tree constructed on the training lexicon is subsequently employed as the core of the letter-phoneme converter of the TREETALK system. To retrieve the phonemization of a word, whether it is in the training lexicon or unseen, all letters of the word are taken as starting points of tree search one by one. With each letter in focus, the search traverses the tree to the point where a leaf node is met, or where the search fails as the specific context of the new word was not encountered in the learning material; in either case a phoneme is returned, stored at the node that was last visited. The complete phonemization of the new word is assembled by concatenating IGTREE's phonetic classifications of each of the word's letters, deleting the null phonemes.

4.1.4 Experiments: applying IGTREE to word phonemization

We base our experiments on a corpus of English word phonemizations, which we extracted from the CELEX-2 lexical databases (Baayen et al., 1993) already mentioned in the previous chapter. The corpus contains 65,467 words for which word phonemization is stored along with word stress information, syllable boundaries, and a morphological analysis.

On the basis of the CELEX-2 corpus of words with their corresponding phonemic transcription we create a database of 573,170 examples, in which each example is a windowed letter-in-context associated with the phoneme that the middle letter maps to (cf. Table 4.1). Fifty-eight unique phonemes occur in the data.

We performed comparative experiments with IB1 and IGTREE, both using gain ratio; IB1 for feature weighting, and IGTREE for feature ordering. IB1 uses the Overlap similarity metric and $k = 1$ (in TIMBL this is the default parameter setting). The database was partitioned ten times into 90% training set and 10% test set parts (each test set is disjoint from every other test set, and the partitions are made on the word level on a randomly shuffled lexicon, so that all words are in a test set once and no alphabetic ordering occurs in training or test data) and a 10-fold cross-validation experiment was run (Weiss & Kulikowski, 1991). Table 4.2 displays the test word accuracy averaged over the ten experiments; a test word is counted as correctly phonemized when its phonemes are all perfectly predicted. The table also displays the "reproductive" accuracy, i.e., the percentage of correctly phonemized training words. This score is below 100%. If a memory-based learner does not classify its own training examples with 100% accuracy, this means there are examples with identical feature values and different class labels. This is due to our choice of context width encoded in the window, that leaves ambiguities open because it is too narrow. For instance, the example generated around the o of collar, _ _ _ c o l l, matches with a set of morphologically complex words starting with coll (collateral, collective, etc.) in which the o is pronounced as [ə]. Most of the observed reproduction errors are incorrect vowel pronunciations due to unsolved ambiguities in which different nearest neighbors have different pronunciations, and correlated to this, different word stress or morphological structures, despite their superficial similarity.

As Table 4.2 shows, there is no accuracy difference between IB1 and IGTREE on training data; not surprisingly, they make the same errors due to the unsolvable ambiguity in the data. On test words, however, IB1 performs better than IGTREE. When phonemizing test words IB1

| | Correctly phonemized words (%) | |
Algorithm	Training set	Test set
IB1	92.1 ± 0.1	80.9 ± 0.4
IGTREE	92.1 ± 0.1	78.5 ± 0.4

Table 4.2: Average generalization performance in terms of correctly phonemized training words (left) and test words (right).

| | | Instances per second | |
Algorithm	Storage (nodes)	Training set	Test set
IB1	628,975	22,737	9,976
IGTREE	31,694	26,204	27,665

Table 4.3: Average memory storage (in numbers of nodes) and the number of classified instances per second, classifying both training instances and test instances, by IB1 and IGTREE, averaged over a 10-fold CV experiment on the phonemization data set.

has an intrinsic advantage because it matches on all features, also the relatively less important features beyond a more important mismatching feature, rather than stopping as soon as a feature mismatches or when the disambiguating subset of features as stored in the IGTREE is already tested.

Yet, there is an interesting trade-off between accuracy on the one hand, and speed and memory on the other hand between IB1 and IGTREE. Table 4.3 displays the average memory storage (in terms of nodes in both algorithm's trees – recall that the instance base of IB1 is also stored as a decision tree) and processing costs (in terms of the number of instances classified per second, on average[1]) of both algorithms. The advantage in generalization accuracy of IB1 over IGTREE is at the cost of considerably more memory storage; IGTREE occupies 95.0% less memory than IB1[2]. While IB1 stores about as many nodes in its tree as there are training examples, IGTREE uses one node per 20 training examples on average. Moreover, IGTREE is about 2.8 times faster than IB1 on test data.

---

[1]Processing times were measured on a PC equipped with a 2.7 Ghz Intel processor running GNU/Linux.

[2]Actual memory usage depends on the platform – on a standard current desktop computer, one node uses 20 bytes.

4.1. PHONEMIZATION

The scores on phonemizing test words (80.9% at best, with IB1) are not impressive, but it may serve as a consolation that at the phoneme level, IB1 predicts 97.0% correct phonemes. Also, this is a phonemization score on unseen words, since they are produced by a 10-fold cross-validation experiment on a type lexicon in which words only occur once. In a realistic setting, e.g. as part of a speech synthesis system phonemizing words in running text, typically about 5% of the words in the text would be unseen. The system would thus be predicting 95% of the words at 92.1% accuracy, and 5% at 80.9%, which amounts to the very reasonable estimated performance levels of 91.5% correctly phonemized words, composed of 98.8% correctly predicted phonemes – well within the aforementioned perceptually significant 5% error threshold on phonemes.

4.1.5 TRIBL: trading memory for speed

The results discussed in the previous section show that, for the phonemization task, there is a notable performance loss when using IGTREE compared to pure memory-based learning as used in IB1 on the word phonemization task. GTREE and IB1 represent extremes on a continuum which would be interesting to explore in a more scaled way. To this purpose we developed TRIBL, a hybrid between IB1 and IGTREE. TRIBL uses a new parameter that determines the relative contribution of IGTREE as opposed to IB1, to learning as well as to classification. The parameter value denotes the last level at which IGTREE compresses homogeneous subsets of instances into leaves or non-terminal nodes; below this level, all remaining uninspected feature values of instances in non-homogeneous subsets, normally represented by a single node in the decision tree, are stored uncompressed in instance-base nodes. During search, the normal IGTREE search algorithm is used for the levels that have been constructed by IGTREE; when IGTREE search has not ended at the level marked as the switch point between IGTREE and IB1, one of the instance-base nodes is accessed, and IBL is employed on the sub-instance-base stored in this instance-base node.

We learned the word phonemization task with TRIBL instead of IGTREE[3]. The empirical results are shown in Table 4.4. Again, the results were achieved by means of 10-fold CV experiments. The table repeats the results obtained with IB1 and IGTREE, and lists in between the

---
[3]See Daelemans et al. (1997c) for a discussion of the results of TRIBL on several non-linguistic benchmark data sets.

Algorithm	Correctly phonemized test words (%)	Instances per second
IB1	80.9 ± 0.4	9,976
TRIBL-1	80.8 ± 0.3	11,257
TRIBL-2	80.7 ± 0.3	14,308
TRIBL-3	80.7 ± 0.3	15,261
TRIBL-4	79.6 ± 0.3	17,643
TRIBL-5	79.0 ± 0.4	21,172
TRIBL-6	78.5 ± 0.4	25,647
IGTREE	78.5 ± 0.4	27,665

Table 4.4: Average percentage of correctly phonemized test words, with standard deviation, of IB1, TRIBL with all possible switch points, and IGTREE, with the average number of test examples classified per second, on the word phonemization data set.

results obtained with different values for the various TRIBL switch points from IGTREE to IB1. We show both accuracy and efficiency in terms of processing time. The results listed in Table 4.4 show that TRIBL offers an interesting non-linear trade-off between processing time on the one hand, and generalization accuracy on the other hand. At TRIBL with splitting at tree level 3, processing speed has almost doubled, while at that point word phonemization accuracy is hardly affected.

On the one hand, TRIBL can be seen as a means to find the "best of both worlds" of IB1 and IGTREE. On the other hand, TRIBL may provide relevant linguistic insights with respect to the relative importance of features, such as local context in sequential NLP tasks. A sequence of results such as displayed in Table 4.4 shows that generalization accuracy remains unchanged when the classifier uses an absolute preference order in matching on the most important features (here, the focus letter and its two immediately neighboring letters). It may be interesting to compare this finding with rule-based approaches to the same task, which may impose the same or different restrictions on the context-sensitivity of their rules.

4.1.6 TRIBL in TIMBL

TRIBL is included in TIMBL. As a hybrid of IB1 and IGTREE, it is governed by the same parameters; the feature weighting parameter determines the feature weighting in the similarity function of IB1 as well as the ranking of the features in the construction of the IGTREE.

The -q parameter determines the number of features (tree levels) after which IGTREE switches to IB1. On the German plural data set used earlier in chapter 3, TRIBL runs as follows, e.g., with q set to 1 (splitting on the feature with the highest GR, gender), and performing IB1 on the remaining six features:

```
% Timbl -f gplural.train -t gplural.test -a2 -q1
```

Setting q to 1 leads to 94.14% correctly classified test instances. Sweeping q from 1 to 6 yields a maximal score of 94.52% with q at 4. This means that the gender feature and the three segmental features of the last syllable (onset, nucleus, coda) are used to build an IGTREE and the other features (the pre-final syllable features) are handled using IB1.

4.2 Morphological analysis

The task of performing a full morphological analysis of a wordform is usually taken as a segmentation of the word into morphemes, combined with an analysis of the interaction of those morphemes that determine the syntactic class of the wordform as a whole. The complexity of wordform morphology varies widely among the world's languages, but is regarded as non-trivial even in the relatively simple cases, such as English. Classes of linguistic knowledge that are usually assumed to play a role in this disambiguation process are knowledge of (i) the morphemes of a language, (ii) the morphotactics, i.e., constraints on how morphemes are allowed to be combined, and (iii) spelling changes in the resulting word form that can occur due to morpheme attachment.

The memory-based approach to morphological analysis (including compounding) of complex wordforms we illustrate here will be referred to from now as MBMA, for Memory-Based Morphological Analysis. We exemplify the approach by turning to the morphological analysis of Dutch words as a case study.

4.2.1 Dutch morphology

Dutch, a descendant of the North-Sea Germanic branch of the Germanic languages, has a non-trivial wordform morphology that shares some of its traits with the wordform morphology of German. The processes of Dutch morphology include inflection, derivation, and compounding. Inflection of verbs, adjectives, and nouns is mostly achieved by suffixation, but a circumfix also occurs in the Dutch past participle (e.g., ge+werk+t as the past participle of verb werken, to work). Irregular inflectional morphology is due to relics of ablaut (vowel change) and to suppletion (mixing of different roots in inflectional paradigms). Processes of derivation in Dutch morphology occur by means of prefixation and suffixation. Derivation can change the syntactic class of wordforms. Compounding in Dutch is concatenative (as in German and in Scandinavian languages): words can be strung together almost unlimitedly, with only a few morphotactic constraints, e.g., rechtsinformaticatoepassingen (applications of computer science in law).

A complex Dutch wordform typically inherits its syntactic properties from its right-most part (the head). Several spelling changes can occur: apart from the closed set of spelling changes due to irregular morphology, a number of spelling changes is predictably due to morphological context. The spelling of long vowels varies between double and single (e.g., ik loop, I run, versus wij lop+en, we run), both of which feature the pronunciation [o]. The spelling of root-final consonants can be doubled (e.g., ik stop, I stop, versus wij stopp+en, we stop). There is variation between s and z and f and v (e.g., huis, house, versus huizen, houses). Finally, between the parts of a compound, a linking morpheme may appear (e.g., staat+s+loterij, state lottery). For a detailed discussion of morphological phenomena in Dutch, see Booij (2001).

4.2.2 Feature and class encoding

We drew our data from the CELEX-2 lexical database collection (Baayen et al., 1993). CELEX-2 offers a morphological analysis for 336,698 of them. We took each wordform and its associated analysis, and created task instances using the windowing method exemplified in the previous section on word phonemization. Windowing transforms each wordform into as

4.2. MORPHOLOGICAL ANALYSIS

Instance number	Left context					Focus letter	Right context					Class
1	_	_	_	_	_	a	b	n	o	r	m	A
2	_	_	_	_	a	b	n	o	r	m	a	0
3	_	_	_	a	b	n	o	r	m	a	l	0
4	_	_	a	b	n	o	r	m	a	l	i	0
5	_	a	b	n	o	r	m	a	l	i	t	0
6	a	b	n	o	r	m	a	l	i	t	e	0
7	b	n	o	r	m	a	l	i	t	e	i	0
8	n	o	r	m	a	l	i	t	e	i	t	0+Da
9	o	r	m	a	l	i	t	e	i	t	e	A_→N
10	r	m	a	l	i	t	e	i	t	e	n	0
11	m	a	l	i	t	e	i	t	e	n	_	0
12	a	l	i	t	e	i	t	e	n	_	_	0
13	l	i	t	e	i	t	e	n	_	_	_	0
14	i	t	e	i	t	e	n	_	_	_	_	plural
15	t	e	i	t	e	n	_	_	_	_	_	0

Table 4.5: Instances with morphological analysis classifications derived from the example word abnormaliteiten.

many instances as it has letters. Each example focuses on one letter, and includes a fixed number of left and right neighbor letters, chosen here to be five. Consequently, each instance spans eleven letters, which also happens to be the average word length in the CELEX-2 database.

To illustrate the construction of instances, Table 4.5 displays the 15 instances derived from the Dutch example word abnormaliteiten (abnormalities) and their associated classes. The class of the first instance is A, which signifies that the morpheme starting in a is an adjective (A). The class of the eighth instance, 0+Da, indicates that at that position no segment starts (0), but that an a was deleted at that position (+Da, "delete a" here). Next to deletions, insertions (+I) and replacements (+R, with a deletion and an insertion argument) can also occur. Together these two classification labels code that the first morpheme is the adjective abnormaal. The second morpheme, the suffix iteit, has class A_→N. This complex tag, which is in fact a rewrite rule, indicates that when iteit attaches right to an adjective (encoded by A_), the new combination becomes a noun (→N). Rewrite rule class labels occur exclusively with suffixes, that do not have a part-of-speech tag of their own, but rather seek an attachment to form a complex morpheme with the part-of-speech tag. Finally, the third morpheme is en, which is a plural inflection that by definition attaches to a noun.

When a wordform is listed in CELEX-2 as having more than one possible morphological labeling (e.g., a morpheme may be N or V, the inflection -en may be plural for nouns or infinitive for verbs), these labels are joined into ambiguous classes (N/V) and the first generated example is labeled with this ambiguous class. Ambiguity in syntactic and inflectional tags occurs in 3.6% of all morphemes in our CELEX-2 data.

Encoding the data this way, we generated a sizable data set of 3,179,383 examples; 2,738 different class labels occur. The most frequently occurring class label is 0, occurring in 68.8% of all instances. The three most frequent non-null labels are N (start of noun stem, 6.9%), V (start of verb stem, 3.6%), and plural (start of plural inflection, 1.6%). Many class labels combine a syntactic or inflectional tag with a spelling change, and generally have a low frequency.

Figure 4.4 illustrates how a morphological analysis is regenerated from the letter-by-letter class labels, using the same example word. This regeneration process enacts the operations encoded in the classes. The first class label marks the start of the morphemic segment abnormal and tags it as A (adjective); the second marks the deletion of the second a before the final l of abnormal. The third marks the start of the suffix iteit and tags it with the rewrite rule $A_- \rightarrow N$. The fourth identifies the plural inflection en. Together, this leads to the flat bracketed analysis $[abnormaal]_A[iteit]_{A_- \rightarrow N}[en]_{plural}$ in the middle of Figure 4.4. Finally, the rewrite rule is rewritten and the plural is attached to the resulting noun, to create the nested analysis $[[[abnormaal]_A$ $iteit]_N \ en]_{plural}$, the intended output of MBMA.

4.2.3 Experiments: MBMA on Dutch wordforms

We performed 10-fold cross validation experiments using the 3.2 million examples data set, using the aforementioned window width of five left and right context letters. Again we apply both IB1 and IGTREE to the task. Both use gain ratio; IB1 uses the Overlap metric and $k = 1$. Table 4.6 lists the average percentage of test words that are morphologically analyzed fully correctly. Both IB1's and IGTREE's reproductive accuracy on training words is notably below 100%. For a large part this is due to several inherent ambiguities in Dutch verbal inflections. Words ending in -de may be past tense singular forms, or adjectival past participles ending in -d with an adjectival -e inflection. The final instances of such words would not only

4.2. MORPHOLOGICAL ANALYSIS

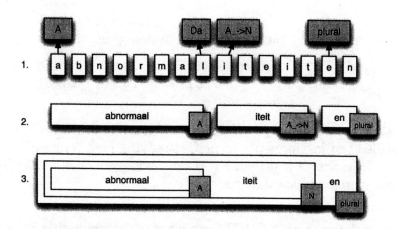

Figure 4.4: Reconstruction of the morphological analysis of the example word abnormaliteiten. Four non-null classes are predicted in the classification step (1); the letters of the three identified morphemic segments are concatenated, the a is inserted, and the syntactic tags are attached to the segments (2); the syntactic rewrite tag of the middle segment is rewritten, and the morphemes are nested (3).

	Correctly analyzed words (%)	
Algorithm	Training set	Test set
IB1	80.5 ± 0.0	61.2 ± 0.2
IGTREE	80.5 ± 0.0	59.1 ± 0.3

Table 4.6: Average generalization performance in terms of correctly morphologically analyzed training words (left) and test words (right).

have themselves as the nearest neighbors in the k-nearest distances, but also several identical examples with conflicting classes. The core problem is that some words with a de ending can only be past tense singulars, such as aaide (caressed), while all adjectival past participles can in principle also be past tense singular forms, such as voltooide (completed) as in het voltooide werk (the completed work) versus hij voltooide het werk (he completed the work). Having an ambiguous class for the two competing forms, and letting a part-of-speech tagger make the decision, would likely be a good solution to this problem, but for now we regard it as an unsolvable ambiguity.

The generalization accuracy of IB1 on completely correctly analyzed test words represents almost a doubling of the error on training material,

Algorithm	Storage (nodes)	Instances per second	
		Training set	Test set
IB1	6,853,002	18,697	1,164
IGTREE	237,571	26,399	24,661

Table 4.7: Average memory storage (in numbers of nodes) and the number of classified test instances per second, classifying both training instances and test instances, by IB1 and IGTREE, averaged over a 10-fold CV experiment on the MBMA data set.

bringing it down to 61.2%. IGTREE's error is slightly higher than IB1's, but its 2.1% loss in accuracy is compensated, as shown in Table 4.7, by a spectacular 96.5% reduction in memory usage, and a speed increase of about factor 21. Remarkably, the speed reported here is close to the classification speed reported for the word phonemization task (cf. Table 4.3), even though the average tree built for this task is an order of magnitude larger – classification in the trees of both tasks occurs at a similar average depth.

We can measure more performance scores than word accuracy, speed, and memory usage. Beneath the word level, analysis performance can be measured at finer granularities. One obvious candidate performance metric would be the average accuracy at the level of classifications. For IB1, for example, this is 94.4% on test data, and 97.6% on training data (so, the 19.5% word error rate is due to a 2.4% classification error rate). These classification accuracy numbers are not informative, however, since the target task, morphological analysis, deviates slightly but essentially from the classification task we redefined it as. The major deviation is that we introduced a "do nothing" class with the label "0". When a simple classifier would always predict "0", being the majority class label with a 68.8% occurrence rate, this classifier would score 68.8% correct classifications, but it would effectively predict 0% of the morphological analyses. Any metric that estimates the generalization performance on morphological analysis should measure the 0%, and not the 68.8%.

To this purpose, as introduced in section 3.5, we can use precision, recall, and their harmonic mean, F-score. Here, precision is the percentage of morphemes identified by MBMA that are indeed morphemes in the target analysis; recall is the percentage of morphemes in the target analysis that are also predicted by MBMA. A high precision signifies that MBMA is

4.2. MORPHOLOGICAL ANALYSIS

able to identify morphemes accurately, but it does not specify how many morphemes it does not identify. Analogously, a high recall signifies that MBMA is able to identify a lot of the morphemes it is supposed to identify, but it does not specify how many more incorrect morphemes it identifies. As mentioned earlier in section 3.5, precision and recall can be merged in a single F-score, which is their harmonic mean. A more generic definition of F-score is $F_\beta = \frac{(\beta^2+1) \cdot \text{precision} \cdot \text{recall}}{\beta^2 \cdot \text{precision} + \text{recall}}$; where β is a means to assign a higher weight either to precision or recall; we set $\beta=1$ to give equal weight to both.

Subsequently, we can analyze the output of MBMA at several levels of granularity. The full task is to identify the boundaries of morphemes along with their part-of-speech tags and possible spelling changes, but simpler subtasks can be discerned within this full task. We provide post-hoc analyses with precision, recall, and F-score at the following levels:

1. Full morphological analysis, in terms of the precision and recall of correctly identified morphemes. A morpheme is correctly identified if its boundaries are correctly identified, the basic part-of-speech tags (including rewrite rules) are correctly placed, and the correct spelling change is predicted.

2. Segmentation: The precision and recall of correctly identified segments. We discern between the following two metrics:

    (a) Typed segmentation. A segment is correctly identified if its boundaries are correctly identified and it is tagged as an inflectional morpheme or a non-inflectional morpheme (i.e., a stem or an affix).

    (b) Plain segmentation. A segment is correctly identified if its boundaries are correctly identified.

Table 4.8 displays these performance metrics for IB1 on test material. The F-score on the full task is 80.9%. If part-of-speech tags and spelling changes are discarded from the classification output, the F-score of both segmentation tasks improves with an error reduction of about 50%. Apparently it is about as hard to segment without types, as it is to segment and identify each segment as an inflection or a non-inflection – the latter subtask appears to be very predictable.

A similar consolation as with the word phonemization task applies here as regards the fact that these performance metrics are based on held-out lexicon types, i.e., typical rare, unseen words, and that the reproduction

Subtask	Correctly identified morphemes (%)		
	Precision	Recall	F-score
Full morphological analysis	81.1 ± 0.1	80.7 ± 0.2	80.9 ± 0.1
Typed segmentation	90.3 ± 0.1	89.9 ± 0.1	90.1 ± 0.1
Untyped segmentation	90.4 ± 0.1	90.0 ± 0.1	90.2 ± 0.1

Table 4.8: Average precision, recall, and F-score of correctly identified morphemes at different granularity levels, analyzed post-hoc from the output on test material of IB1 on the MBMA task.

performance on known words is higher. Counting with the same 95% known words – 5% unseen words free-text estimate, the word accuracy of IB1 would be 79.5%. Likewise, the F-score on fully correctly identified morphemes (including part-of-speech tag and spelling change) would be 90.6%; typed and untyped segmentation would have respectable F-scores around 96.0%.

Morphological analysis is tightly coupled with part-of-speech tagging, which is covered in the next chapter. Obviously, MBMA is already performing part-of-speech tagging, but without information on the words surrounding the analyzed word. It is already doing a good job as a part-of-speech tagger; it correctly predicts the main tag (out of the 13 main tags discerned by CELEX-2; noun, verb, adjective, etc.) of 87.1% of the unseen wordforms, and 94.0% of the known words – a free-text estimate of 93.7% correct tags. An external part-of-speech tagger could benefit from MBMA's suggestion by incorporating the (possibly still ambiguous) tag predicted by MBMA in its own tagging decision which is typically based on context rather than on an analysis of the focus word itself.

4.3 Conclusion

In this chapter we showed that English word phonemization and Dutch morphological analysis can be formulated as classification tasks. Word phonemization is formulated as a disambiguation task, in which one letter in context is to be mapped to its corresponding phonemic transliteration. Morphological analysis is formulated as a complex segmentation task; at segmentational boundaries, the class label marking the segmentation also carries information about spelling changes and part-of-speech information.

4.3. CONCLUSION

We presented memory-based models of the two tasks, showing how columns in a lexical database such as CELEX-2 can be converted into a word phonemization system or a morphological analyzer, which at the same time are able to reproduce the information in the lexicon, and are able to generalize to unseen words. We reported on generalization performances of both models on unseen test words, and compared those of IB1 with IGTREE, a fast decision-tree approximation of the normal k-NN classifier that typically loses a bit of accuracy, but does so with markable speed gains and memory usage reductions.

The results indicate that IGTREE is a useful variant of memory-based learning. With IGTREE we are able to produce an English word phonemization system of which the decision tree fits in about 634 kilobytes of memory (counting with 31,694 nodes and a storage cost of 20 bytes per node), which is able to phonemize about five thousand words per second (counting with an average word length of five in normal English texts, and a processing speed of about 27,000 instances per second), at a performance level that is well within bounds of the minimal requirements for being useful in a speech synthesis system. As said, Yvon (1996) claims that at least 95% of all phonemes must be predicted correctly to produce perceptually acceptable synthesized speech; we estimated that on running text our system would predict 98.8% of all phonemes correctly, which is certainly accurate enough. This combination of accuracy, high speed, and moderate memory requirements makes the TREETALK system very suited for integration in speech synthesizers.[4]

The Dutch morphological analysis system, trained on about three million windowed examples, is able to analyze about five thousand words per second as well, using about 4.6 megabytes of memory. The performance measured at the level of full analysis (segmentation, spelling changes, and part-of-speech information) is somewhat low (59.1% correctly analyzed words, at an F-score of 80.9% of correctly analyzed morphemes), but when measured at the level of correct segmentation, a high F-score of 90.2% on unseen words is obtained. For a morphologically productive language such as Dutch it is important to have high-quality morphological analysis for use in higher-level tasks such as information retrieval, to use the compounding morphemes as individual index terms. This has been shown to boost both precision and recall, also if the compounds are simply segmented, and CELEX-2 is only used for lookup (Pohlmann & Kraaij,

---

[4] See, for example, the NeXTeNS speech synthesis system for Dutch, http://nextens.uvt.nl.

1997).

The speeds mentioned for the two systems are obtained with IGTREE; at much slower speeds, IB1 is able to attain somewhat higher generalization accuracies. This shows that IGTREE can be a very effective approximation of IB1, but it also shows that the absolute feature ordering of IGTREE does not capture the tasks as well as the feature weighting in IB1 does. In addition, we demonstrated that with the TRIBL algorithm the continuum between IB1 and IGTREE can be explored; it appears possible to attain considerable speed gains with hardly any performance loss when tuning TRIBL correctly.

An important problem with the two example systems lies in the use of the windowing method to generate examples with single class labels. A classifier that is processing a word, letter by letter, is presented with instances that do not reflect any previous or future decisions by the classifier. The classifier could never be aware of dependencies in the output sequence that span over larger distances than the window width. Still, in both tasks such long-distance dependencies exist. To make a difference between the phonemizations of abnormal and abnormality, the word phonemization module should be aware of the ity suffix when determining the pronunciation of the first a and o. Likewise, a typical verb inflection at the end of a long, unseen word should be a clue that the whole word is a verb. Still, neither of the systems described in this chapter can recognize these clues.

There are at least two solutions. First, the word phonemization system could be equipped with the output of a morphological analyzer in its input. Van den Bosch (1997) explores this modularity solution as part of the development of an English unknown-words word phonemization system. In this study it was observed that a perfect morphological analyzer could help the letter-phoneme conversion, but a trained morphological analyzer produced too many errors to be of help to the letter-phoneme module.

A second solution is presented in chapter 7, in which we show that we can make the classifier aware of its own classifications, and in which we force the classifier to predict n-grams of class labels rather than just a single label. When applied in combination, an error reduction of 34% of the F-score on the full morphological identification task can be obtained.

4.4 Further reading

The earliest applications of machine learning to natural language occurred largely in computational phonology. Among these, MBRTalk (Stanfill & Waltz, 1986; Stanfill, 1987) was the first application of the k-NN classifier approach to the word phonemization task. MBRTalk was trained and tested on the "Nettalk" data set of English word phonemization, developed for a classic experiment with multi-layer perceptrons trained with the error back-propagation learning rule (Sejnowski & Rosenberg, 1986) that for a while was the key example of the applicability of artificial neural networks to NLP tasks. As the "Nettalk" multi-layered perceptron did, MBRTalk used windowing to generate examples; each instance contained one letter to be transcribed, its four left and right neighbor letters, and the transcription of the middle letter: a phoneme with a stress marker indicating whether the phoneme received word stress. MBRTalk employed the Overlap metric without feature weighting. Weijters (1991) introduced a hand-set feature weighting metric in a k-NN classifier with the Overlap metric, trained yet again on the Nettalk data. Van den Bosch and Daelemans (1993) introduced information-gain feature weighting in the Overlap metric (cf. subsection 3.2.1). Decision trees for phonemization have been introduced by Lucassen and Mercer (1984), who also used an information-theoretic metric as a guiding principle in constructing the tree. Their context features are preceding and following letters and preceding phonemes, coded as binary features, which necessitates a recursive search for the most informative binary features using mutual information. The application of Machine Learning algorithms such as C4.5 (Quinlan, 1993) to phonemization is analogous to this approach (see e.g., Dietterich et al., 1995; Ling & Wang, 1996).

The IGTREE approach, although developed independently with a completely different motivation, is functionally quite close to Kohonen's (1986) Dynamically Expanding Context approach (DEC), applied to phonemization in Torkkola (1993). DEC extracts rules from the data according to a handcrafted specificity hierarchy (in this case, specificity of context and the intuition that context further away from the focus position is decreasingly relevant). This is effectively equivalent to IGTREE, but IGTREE computes the specificity hierarchy (the feature ordering) automatically.

While our approach focuses on classification of single letters in context to their phonemic mapping, other approaches have been proposed that map variable-width chunks of letters to chunks of phonemes by analogy (Sullivan & Damper, 1992; Sullivan & Damper, 1993; Pirelli & Federici,

1994; Yvon, 1996; Damper, 1995; Damper & Eastmond, 1997). Many of these systems claim to be motivated by and build further on Glushko's (1979) psycholinguistically oriented single-route model of reading aloud, and Dedina and Nusbaum's (1991) PRONOUNCE model for chunk-based text-to-speech conversion.

Computational morphology (Sproat, 1992) has seen relatively few applications of machine learning methods to the automatic analysis of words. We argued that our work on the morphological analysis of Dutch (Van den Bosch & Daelemans, 1999, section 4.2) and English (Van den Bosch et al., 1996; Van den Bosch, 1997) constituted a single-level morphological analysis engine, to be seen in contrast with the predominant two-level morphology machines using finite-state transducers (Koskenniemi, 1983; Koskenniemi, 1984). Where our approach is shown to work only for analysis, two-level finite state morphological analyzers are bi-directional: they can both analyze and generate. Our approach can only analyze by classification – another classification module is needed to perform generation. The division of labor in two-level finite state transducers is usually between concatenation processing and wordform spelling alternation processing – which in our approach is integrated in one step. See De Pauw et al. (2004) for a comparison of memory-based and finite-state approaches for Dutch. More recently, Clark (2002) pointed out that single-step memory-based morphological analysis still needs a second processing step which implements the complex class labels to generate the analysis, even though this second step is simple. Clark proposes a more direct memory-based model that uses stochastic transducers operating directly on the input string. Heemskerk and Van Heuven (1993) describe a two-level approach to Dutch morphology using context-free word grammars interleaved with exploration of possible spelling changes; Heemskerk (1993) describes a probabilistic variant. This knowledge-based work draws from descriptive linguistic work on Dutch morphology (De Haas & Trommelen, 1993; Booij, 2001).

Published examples of applications of memory-based approaches to other word-level tasks in the morpho-phonological domain are hyphenation and syllabification (Daelemans & Van den Bosch, 1992); assignment of word stress in Dutch (Daelemans et al., 1994); word phonemization in Dutch and French (Van den Bosch & Daelemans, 1993; Busser, 1998); predicting linking morphemes in Dutch compounds (Krott et al., 2001); diminutive formation of Dutch nouns (Daelemans et al., 1998); plural formation of German nouns (Daelemans, 2002); and Spanish stress assignment and English past tense learning (Eddington, 2003).

Chapter 5

Application to shallow parsing

The goal of this chapter is to show that even complex recursive NLP tasks such as parsing (assigning syntactic structure to sentences using a grammar, a lexicon and a search algorithm) can be redefined as a set of cascaded classification problems with separate classifiers for tagging, chunk boundary detection, chunk labeling, relation finding, etc. In such an approach, input vectors represent a focus item and its surrounding context, and output classes represent either a label of the focus (e.g., part of speech tag, constituent label, type of grammatical relation) or a segmentation label (e.g., start or end of a constituent). In this chapter, we show how a shallow parser can be constructed as a cascade of MBLP-classifiers and introduce software that can be used for the development of memory-based taggers and chunkers.

Although in principle full parsing could be achieved in this modular, classification-based way (see section 5.5), this approach is more suited for *shallow parsing*. Partial or shallow parsing, as opposed to full parsing, recovers only a limited amount of syntactic information from natural language sentences. Especially in applications such as information retrieval, question answering, and information extraction, where large volumes of, often ungrammatical, text have to be analyzed in an efficient and robust way, shallow parsing is useful. For these applications a complete syntactic analysis may provide too much or too little information. For example, in text mining applications such as information extraction, summarization, ontology extraction from text and question answering we are more interested in finding concepts (e.g., simple NPs and VPs) and grammatical relations between their heads (e.g., who did what to whom, when, where, why and how) than in elaborate configurational syntactic analyses. Shallow parsing is also useful for reducing the search space of full parsers.

Abney (1991) was the first to argue for the relevance of shallow parsing, both from the point of view of psycholinguistic evidence and from the point of view of practical applications. His own approach used hand-crafted cascaded finite-state transducers to construct a shallow parse. Typical modules within a shallow parser architecture include the following:

1. Part-of-speech (POS) tagging. Given a word and its context, decide what the correct morpho-syntactic class of that word is (noun, verb, etc.). POS tagging is a well-understood problem in NLP (Van Halteren, 1999).

2. Chunking. Given the words and their morpho-syntactic class, decide which words can be grouped as chunks (noun phrases, verb phrases, prepositional phrases, complete clauses, etc.) and determine their heads.

3. Relation finding. Given the NP chunks in a sentence, decide which relations their heads have with the main verb (subject, object, location, etc.).

The concept of shallow parsing has no clearly defined meaning however, and is used sometimes in a very limited sense, referring only to tagging and chunking, and sometimes in a broader sense, referring also to tasks such as prepositional phrase assignment (see section 1.2) and named-entity recognition. It can best be interpreted as a family of related tasks attempting to recover some syntactic-semantic information in a robust and deterministic way at the expense of ignoring detailed configurational syntactic information. In this chapter, we restrict its meaning to the three tasks above and demonstrate an MBLP approach to them. The approach is evaluated on the WSJ treebank corpus (Marcus et al., 1993). We also introduce new software, MBTG and MBT, two wrappers around TIMBL that are useful for tagging and chunking.

5.1 Part-of-speech tagging

Part-of-speech (POS) tagging is a process in which a morpho-syntactic class is assigned to each word in a text on the basis of the word's formal and lexical properties and of the context in which it occurs. It is a first

5.1. PART-OF-SPEECH TAGGING

level of abstraction in text analysis, often used as a preprocessing module in many language technology applications such as parsing, information retrieval, spelling error correction, speech synthesis, and text mining. Just as it is a reliable heuristic in morpho-phonology (see chapter 4) to assume that a spelling symbol will have the same pronunciation or morphological structure decision in similar contexts, the main idea in a memory-based approach to POS tagging is that an ambiguous word will have the same POS tag in similar contexts (Daelemans et al., 1996).

5.1.1 Memory-based tagger architecture

The construction of a POS tagger for a specific corpus is achieved in the following way. Given an annotated corpus, three data structures are automatically extracted: a *lexicon*, an instance base for *known words* (words occurring in the lexicon), and an instance base for *unknown words*. The lexicon associates words with their ambiguous tag, henceforth referred to as *ambitag*: a symbol representing all the POS tags a word can have according to the corpus. E.g., for a word like executive which occurs in the WSJ corpus both as an adjective (JJ) and as a noun (NN), the corresponding lexical ambitag is NN-JJ (the word occurs more frequently as NN than as JJ, hence the order). A word like current also occurs as both JJ and NN, but less as NN, and will therefore get the ambitag JJ-NN.

During tagging, each word in the text to be tagged is looked up in the lexicon. If it is found, its lexical representation is retrieved and its context in the sentence is determined, and the resulting pattern is disambiguated using extrapolation from nearest neighbors in the known words instance base. When a word is not found in the lexicon, its lexical representation is computed on the basis of its form, its context is determined, and the resulting pattern is disambiguated using extrapolation from nearest neighbors in the unknown words instance base. In each case, the output is a best guess of the POS tag for the word in its current context.

The instances are represented by a variety of features of the focus word to be tagged and word forms in its immediate context. The reason for separating known and unknown words is the following: for known words the ambitag of the focus word turns out to be the most important feature. However, for unknown words we do not know the ambitag, and therefore we are restricted to context and word form features to construct the unknown word's instance representation. Below we will use the following notation for the features. Since we go from left to right, we can assume that the words to the left of the word to be tagged have been disambiguated already. These tags are denoted by D, the position of the (ambitag) of the focus word is given by F, and the not yet disambiguated words to the right are denoted by their ambitag A. In both known and unknown word tagging an important source of information is the inclusion of *previous tagger decisions* as features for current tagger decisions with the D features in both known and unknown word tagging. These features allow the approach to escape from the local windowing limitations. Other solutions to sequence learning problems are introduced in chapter 7.

Features referring to particular word forms are denoted as W or W for the word corresponding to the focus position. As the presence of features with thousands of words as feature values would make the tagging considerably slower and low frequency word values would not be likely to match anyway, only the most frequent words (e.g., the 100 most fréquent words) are kept as values, and the others are substituted by the symbol 'HAPAX' annotated with some additional information. E.g., HAPAX-N means that the word contains numeric symbols, HAPAX-C means the word is capitalized, HAPAX-H that it is hyphenated, and HAPAX-0 means no special attributes are associated with the word. A word such as B-52 would then get the value HAPAX-CHN. This process is known as attenuation (Eisner, 1996).

Returning to the level of features rather than values, especially for the unknown word instances there are a number of features referring to the parts of the word form: its suffix letters 'S', prefix letters 'P', a capitalization feature 'C', the presence of a hyphen 'H', and the presence of numerals 'N'. These features provide a kind of "poor man's morphology" that may be useful to guess the POS tag of an unknown word.

Tables 5.1 and 5.2 display example instances from the known words and the unknown words instance bases (on WSJ material) respectively. For the selection of instances for the unknown words case base, only words are selected that occur with relatively low frequency, as these words will have characteristics more similar to unknown words than frequent words.

5.1.2 Results

In previous work on an MBLP approach to tagging (Daelemans & Van den Bosch, 1996; Zavrel & Daelemans, 1997; Van Halteren et al., 2001) for different corpora and different languages, the approach consistently outperforms the well-known transformation-based learning approach (Brill, 1994) and some trigram-based approaches, but often achieves slightly worse results than maximum-entropy approaches (Ratnaparkhi, 1996) and

5.1. PART-OF-SPEECH TAGGING

					Instance representation				
Word	D	W	D	W	F	W	A	W	Class
Consumers	==	==	==	==	NNS	HAPAX-C	MD	HAPAX-0	NNS
may	==	==	NNS	HAPAX-C	MD	HAPAX-0	VBP-VB	HAPAX-0	MD
want	NNS	HAPAX-C	MD	HAPAX-0	VBP-VB	HAPAX-0	TO	to	VB
to	MD	HAPAX-0	VB	HAPAX-0	TO	to	NN-VB-VBP	HAPAX-0	TO
move	VB	HAPAX-0	TO	to	NN-VB-VBP	HAPAX-0	PRP$	their	VB
their	TO	to	VB	HAPAX-0	PRP$	their	NNS	HAPAX-0	PRP$
telephones	VB	HAPAX-0	PRP$	their	NNS	HAPAX-0	DT	a	NNS
a	PRP$	their	NNS	HAPAX-0	DT	a	JJ-RB	HAPAX-0	DT
little	NNS	HAPAX-0	DT	a	JJ-RB	HAPAX-0	RBR	HAPAX-0	RB
closer	DT	a	RB	HAPAX-0	RBR	HAPAX-0	TO	to	RBR
to	RB	HAPAX-0	RBR	HAPAX-0	TO	to	DT	the	TO
the	RBR	HAPAX-0	TO	to	DT	the	NN-NNP	HAPAX-C	DT
TV	TO	to	DT	the	NN-NNP	HAPAX-C	VBN-VB-NN-VBD	HAPAX-0	NN
set	DT	the	NN	HAPAX-C	VBN-VB-NN-VBD	HAPAX-0	.	.	NN
.	NN	HAPAX-C	NN	HAPAX-0	.	.	==	==	.

Table 5.1: Example of instances of the POS tagging task (known words instances). Instances represent fixed-sized snapshots of a focus (an ambitag), surrounded by a left and right context (of disambiguated tags on the left, and ambiguous tags on the right, and highly-frequent words or else attenuated symbols (HAPAX) to the left and to the right).

an HMM-approach with powerful smoothing such as TnT (Brants, 2000). The memory-based approach yields 96.4% accuracy on the WSJ corpus and 97% on the LOB corpus (Zavrel & Daelemans, 1999). Comparison is difficult because of the different data sets, features, and representations used in different learning approaches. Given the minimal language engineering involved in MBLP (making a tagger on the basis of a new annotated corpus is a matter of hours) and the computational efficiency of the method both in training and testing (in the order of thousands of words per second), this state-of-the-art performance is remarkable. In contrast to explicitly probabilistic methods, there is no need in an MBLP

				Instance representation					
Word	P	D	W	A	W	S	S	S	Class
Consumers	C	==	==	MD	HAPAX-0	e	r	s	NNS
television	t	PRP$	their	DT	a	n	e	s	NNS
closer	c	RB	HAPAX-0	TO	to	s	e	r	RBR

Table 5.2: Example of instances of the POS tagging task (unknown words instance base). Instances represent 'morphological' information about the focus word (first letter and the three last letters), surrounded by a left and right context (of one disambiguated tag to the left, one ambiguous tag to the right, and the corresponding attenuated words).

approach for an additional smoothing component for sparse data, as this is already embodied in the similarity-based extrapolation itself (Zavrel & Daelemans, 1997). The use of the weighted similarity metric allows for an easy integration of different information sources (e.g., context tags, words, morphology, spelling, etc.) with no clear a-priori back-off ordering. Moreover, the fact that only one parameter is needed per feature (i.e., its information-theoretic weight) makes MBLP more robust to overfitting than approaches that use very large numbers of parameters. The downside of this robustness is that the feature-weighting capabilities are quite rough: each feature is weighted in isolation, so that no specific weights are assigned to interesting feature combinations, and the weight estimate of conjunctions of redundant features tends to be too large, and there is also no separate weight for specific values of a feature.

5.1.3 Memory-based tagging with MBT and MBTG

MBTG and MBT are two programs built around TIMBL that allow you to construct a tagger on the basis of a tagged corpus (MBTG) and use this tagger to tag new text (MBT). Although in principle it would be possible to use TIMBL directly, the software provides an easy solution to building the different instance bases with relevant word and context features, including preceding tagger decisions. As an example dataset we derived POS information from the CoNLL shared task data (Tjong Kim Sang & Buchholz, 2000), which is a part of the WSJ corpus. See also chapter 6 for a description of these data.

5.1. PART-OF-SPEECH TAGGING

The input file containing the material for generating a tagger must consist of two whitespace-separated columns. The first column contains a word or punctuation mark, as well as its POS tag in the corresponding position of the second column. A line may also contain only the symbol <utt> to mark the end of a sentence. The following is example input.

```
He          PRP
reckons     VBZ
the         DT
current     JJ
account     NN
deficit     NN
will        MD
narrow      VB
to          TO
only        RB
#           #
1.8         CD
billion     CD
in          IN
September   NNP
.           .
<utt>
```

In generating the tagger, information has to be provided to the tagger generator about the context and the form of the words to be tagged. This is done by the parameters -p (feature pattern for known words), and -P (feature pattern for unknown words). Patterns are built up as combinations of the symbols introduced earlier:

For -p and -P

- d   Predicted left context (tag)
- a   Right context (ambitag)
- w   Left or right context (word)
- c   The focus contains capitalized characters
- h   The focus word contains a hyphen
- n   The focus word contains numerical characters
- p   Character at the start of the word
- s   Character at the end of the word

For -p only (known words)

- f   Focus (ambitag for known words)
- W   Focus (word)

For -P only (unknown words)

F   Focus (position of the unknown word)

The symbols d, a, w, p, and s can occur more than once to indicate the scope of the context. Symbols to the left of the focus symbols indicate left context, and symbols to the right of the focus symbols indicate right context.

For example, for known words, the following are a few possible patterns:

dfa	focus ambitag with one disambiguated tag on the left and one ambitag to the right
ddfa	focus ambitag with two disambiguated tags to the left and one ambitag to the right
ddfWa	as previous, plus the focus word (note that W can be declared only immediately after f)
dwdwfWaw	as previous, plus for each context tag the corresponding word (two left, one right)

For unknown words:

dFa	one disambiguated tag to the left and one ambitag to the right
psdFa	as previous, plus the first and last letter of the unknown word to be tagged
psssdFa	as previous, plus the three last letters of the word to be tagged
psssdwFaw	as previous, plus the left and right neighboring words

In addition to constructors for these commonly used features, the MBTG software also allows you to add your own additional features to the instances created using the option -E in combination with adding extra columns to the input file, where each column corresponds to an additional feature associated with the word at that position when used as a focus. An example command line using a file T.train with the POS tagging training data of the CoNLL shared data for tagger generation and corresponding output is the following ("%" is the command line prompt):

5.1. PART-OF-SPEECH TAGGING

```
% Mbtg -p dwdwfWaw -P dwFawpssschn -T T.train

Memory Based Tagger Generator Version 2.0
   (c) ILK and CNTS 1998 - 2004.
   Induction of Linguistic Knowledge Research Group, Tilburg University
   Centre for Dutch Language and Speech, University of Antwerp

   Based on Timbl version 5.1.0 (Release)

Constructing a tagger from: T.train
   Creating lexicon: T.train.lex of 19122 entries.
   Creating ambitag lexicon: T.train.lex.ambi.05
   Creating ambitag translation table: T.train.ambi.05
   Creating list of most frequent words: T.train.top100

   Create known words case base
      Timbl options: ' -a IGTREE +vS  -H'
      Algorithm = IGTREE

      Processing data from the file
      T.train......................................................
         ready: 211727 words processed.
      Creating case base: T.train.known.dwdwfWaw
      Deleted intermediate file: T.train.known.inst.dwdwfWaw
   Create unknown words case base
      Timbl options: ' -a IB1 +vS  -H'
      Algorithm = IB1

      Processing data from the file
      T.train......................................................
         ready: 211727 words processed.
      Creating case base: T.train.unknown.pssschndwFaw
      Deleted intermediate file: T.train.unknown.inst.pssschndwFaw

   Created settings file 'T.train.settings'
Ready:
   Time used: 50
   Words/sec: 4234
```

The output shows which version of TIMBL was used and reports on the generation of a number of data files that will be used by the tagger MBT. These data include the following:

- A frequency-sorted lexicon T.train.lex containing for each word the different tags it was assigned, along with their frequency in the training corpus.
- A lexicon associating with each word an ambitag, derived from the previous lexicon file, and a translation table for associating the generated ambitag letter codes with a more understandable representation. Limited frequency-based smoothing is implemented in this approach: whenever a word–tag combination occurs less than a given percentage (5% by default) of the word's total frequency, it is not included in the ambitag. The parameter -% *<percentage>* modifies this threshold.

- A list with the (by default) 100 most frequent words in the corpus. Only words in this list will be used when the symbols *w*, *W* are used in the -p, -P patterns. The number of most frequent words can be modified with the parameter -M < *number* >. All words *not* in the most-frequent-words list are transformed into the special HAPAX-symbols discussed earlier.
- An instance base for known words. The process consists of two steps. First, instances are created using the specified information sources for known words (as indicated in -p), then the case base is generated from that (which may imply a significant storage reduction, depending on the TIMBL options used, in this case IGTREE). Finally, the intermediate file with instances is deleted — this can be overruled with the option -X.
- An instance base for unknown words. It is parallel to the procedure for known words, but it uses information sources specified in the -P pattern, and uses as default TIMBL settings the IB1-IG algorithm (which uses the overlap metric with gain ratio feature weighting).
- The tagger generation process ends with some information about the real time needed to construct the tagger (total time used and number of words per second), and with the construction of a settings file, which will be used by the MBT executable to use the tagger on new data.

The settings for our training data are the following:

```
e <utt>
l T.train.lex.ambi.05
k T.train.known.ddfa
u T.train.unknown.dFapsss
r T.train.ambi.05
p ddfa
P dFapsss
O K: -a IGTREE +vS U: -a IB1 +vS
L T.train.top100
```

Given that Mbtg was used to generate data files and a settings file defining a memory-based tagger, Mbt can be used to tag text. For example, continuing our example:

5.1. PART-OF-SPEECH TAGGING

```
% Mbt -T T.test -s T.train.settings

Memory Based Tagger Version 2.0
   (c) ILK and CNTS 1998 - 2004.
   Induction of Linguistic Knowledge Research Group, Tilburg University
   Centre for Dutch Language and Speech, University of Antwerp

Based on Timbl version 5.1.0 (Release)
Reading the ambitags from: T.train.ambi.05...ready, (246 tags).
Reading the lexicon from: T.train.lex.ambi.05...ready, (19122 words).
Reading frequent words list from: T.train.top100...ready, (100 words).
Reading case-base for known words from:
T.train.known.dwdwfWaw... ready.
Reading case-base for unknown words from:
T.train.unknown.pssschndwFaw... ready.
Sentence delimiter set to '<utt>'
Beam size = 1
Known Tree, Algorithm = IGTREE
Unknown Tree, Algorithm = IB1

Processing data from the file T.test: ....................

Rockwell          //      NNP     NNP
International     /       NNP     NNP
Corp.             /       NNP     NNP
's                /       POS     POS
Tulsa             /       NNP     NNP
unit              /       NN      NN
said              /       VBD     VBD
it                /       PRP     PRP
signed            /       VBD     VBD
...

Done:
47377 words processed.
 Known    words: 42131  correct from 44075 (95.5893 %)
 Unknown  words: 2678   correct from 3302  (81.1024 %)
 Total         : 44809  correct from 47377 (94.5797 %)
 Time used: 41
 Words/sec: 1155
```

Calling MBT with the settings file of the trained tagger starts the memory-based tagger by reading the data files and a test input file (in this case in the same format as the training data), and sends the tagged input to standard output computing accuracy statistics by comparing the predicted tags to the gold standard ones provided in the test file. The tagger can also read untagged text from input or from a text file. The text should then be tokenized (i.e., punctuation marks should be separated from the words). In the output, word and predicted tag are separated by a single slash (known word) or a double slash (unknown word).

More parameters are available to modify the behavior of the MBTG and MBT executables and to use the software in client-server mode, for those we refer to the reference guide accompanying the software (Daelemans et al., 2003).

5.2 Constituent chunking

As soon as sentences have been disambiguated at the word level concerning their morpho-syntactic category, a next step in shallow parsing will group words into phrases and assign a label to these phrases. If we restrict chunking to finding non-overlapping and non-recursive base chunks, the task can be defined as a classification task by generalizing the approach of Ramshaw and Marcus (1995), who proposed to convert NP-chunking to tagging each word with **I** for a word inside an NP, **O** for outside an NP, and **B** for the start of an NP that is preceded by another NP. The decision on these so-called IOB tags for a word can be made by looking at the POS tag and the identity of the focus word and its local context. For the more general task of chunking other non-recursive phrases, we simply extended the tag set with IOB tags for each type of phrase. To illustrate this encoding with the extended IOB tag set, we can represent the following tagged and chunked sentence:

But/CC [NP the/DT dollar/NN NP] [ADVP later/RB ADVP] [VP rebounded/VBD VP] ,/, [VP finishing/VBG VP] [ADJP slightly/RB higher/R ADJP] [Prep against/IN Prep] [NP the/DT yen/NNS NP] [ADJP although/IN ADJP] [ADJP slightly/RB lower/JJR ADJP] [Prep against/IN Prep] [NP the/DT mark/NN NP] ./.

as:

But/CC$_O$ the/DT$_{I-NP}$ dollar/NN$_{I-NP}$ later/RB$_{I-ADVP}$ rebounded/VBD$_{I-VP}$,/,$_O$ finishing/VBG$_{I-VP}$ slightly/RB$_{I-ADVP}$ higher/RBR$_{I-ADVP}$ against/IN$_{I-Prep}$ the/DT$_{I-NP}$ yen/NNS$_{I-NP}$ although/IN$_{I-ADJP}$ slightly/RB$_{B-ADJP}$ lower/JJR$_{I-ADJP}$ against/IN$_{I-Prep}$ the/DT$_{I-NP}$ mark/NN$_{I-NP}$ ./.$_O$

This representation can then be used to generate instances using a moving window approach exactly the same way as is done in POS tagging.

5.2.1 Results

Table 5.3 (from Buchholz et al., 1999) shows the accuracy of this memory-based chunking approach when training and testing on Wall Street Journal material. We report on precision, recall, and F-scores (with $\beta = 1$). In this case, the features for the MBLP-classifier are the word form and the POS tag as provided by the tagger of the two words to the left, the focus word, and one word to the right (Veenstra, 1998; Tjong Kim Sang & Veenstra, 1999). Adverbial functions are included here as chunking results as well: this classifier assigns adverbial functions such as *locative* or *temporal* to the chunks.

5.2. CONSTITUENT CHUNKING

Type	Precision	Recall	F-score
NPchunks	92.5	92.2	92.3
VPchunks	91.9	91.7	91.8
ADJPchunks	68.4	65.0	66.7
ADVPchunks	78.0	77.9	77.9
PPchunks	91.9	92.2	92.0
ADVFUNCs	78.0	69.5	73.5

Table 5.3: Results of chunking (in %): segmentation and labeling experiments. Reproduced from (Buchholz et al., 1999).

5.2.2 Using MBT and MBTG for chunking

Although we could easily use TIMBL for chunking by combining words and output of the POS tagger to construct the features of the instances, as was done in the work reported earlier in this chapter, an alternative approach is to use the convenience of MBTG and MBT to construct a combined tagger-chunker. This can be achieved by concatenating the POS-tag and the IOB-tag associated with each word in the sentence, and using MBTG to construct a tagger for these combined tags. The following is an example of the type of input file needed for this.

```
He        PRP/B-NP
reckons   VBZ/B-VP
the       DT/B-NP
current   JJ/I-NP
account   NN/I-NP
deficit   NN/I-NP
will      MD/B-VP
narrow    VB/I-VP
to        TO/B-PP
only      RB/B-NP
#         #/I-NP
1.8       CD/I-NP
billion   CD/I-NP
in        IN/B-PP
September NNP/B-NP
.         ./O
<utt>
```

APPLICATION TO SHALLOW PARSING

As POS tagging and chunking are very much related, it makes sense to combine these two learning tasks into one. Interestingly, although the learning task becomes more complex (more classes have to be learned that describe a more complex output space) and data sparseness therefore increases, the results on tagging and chunking separately do not degrade. The following shows the output and the results of the tagger generation and tagging phases using this combined approach (without any specific optimization for this task).

```
% Mbtg -T TC.train -p dwdwfWaw -P pssschndwFaw
...
   Created settings file 'TC.train.settings'
Ready:
   Time used: 55
   Words/sec: 3849

% Mbt -s TC.train.settings -T TC.test
Memory Based Tagger Version 2.0
   (c) ILK and CNTS 1998 - 2004.
   Induction of Linguistic Knowledge Research Group, Tilburg University
   Centre for Dutch Language and Speech, University of Antwerp

Based on Timbl version 5.1.0 (release)

Reading the ambitags from: TC.train.ambi.05...ready, (1468 tags).
Reading the lexicon from: TC.train.lex.ambi.05...ready, (19122 words).
Reading frequent words list from: TC.train.top100...ready, (100 words).
Reading case-base for known words from:
TC.train.known.dwdwfWaw... ready.
Reading case-base for unknown words from:
TC.train.unknown.pssschndwFaw... ready.
Sentence delimiter set to '<utt>'
Beam size = 1
Known Tree, Algorithm = IGTREE
Unknown Tree, Algorithm = IB1
Processing data from the file TC.test: ...................

Rockwell            //      NNP/B-NP        NNP/B-NP
International       /       NNP/I-NP        NNP/I-NP
Corp.               /       NNP/I-NP        NNP/I-NP
's                  /       POS/B-NP        POS/B-NP
Tulsa               /       NNP/I-NP        NNP/I-NP
unit                /       NN/I-NP         NN/I-NP
said                /       VBD/B-VP        VBD/B-VP
it                  /       PRP/B-NP        PRP/B-NP
signed              /       VBD/B-VP        VBD/B-VP
...

Done:
   47377 words processed.
   Known   words: 39226    correct from 44075 (88.9983 %)
   Unknown words: 2443     correct from 3302  (73.9855 %)
   Total        : 41669    correct from 47377 (87.952 %)
   Time used: 49
   Words/sec: 966
```

5.3. RELATION FINDING

Tagging accuracy drops only slightly to 94.5% (from 94.6%) compared to the tagger trained uniquely on POS tagging information. The precision, recall and F-score for chunking over all constituent types are 83.5%, 87.2%, and 85.3%, respectively. In the following chapters we will return to this sequence learning task with approaches providing better results.

5.3 Relation finding

After POS tagging, phrase chunking and labeling, the last step of shallow parsing consists of resolving the attachment between labeled phrases. Work on an MBLP approach to relation finding evolved from work on complement-adjunct distinction (Buchholz, 1998) to subject and object detection (Daelemans et al., 1999), and, finally, relation finding for all relations annotated in the WSJ corpus (Buchholz et al., 1999; Buchholz, 2002).

In this approach, relation finding is done by using a classifier to assign a grammatical relation (GR) between pairs of words in a sentence. One of these words is always a verb, since this yields the most important GRs. The other word (the focus) is the head of a phrase that can be assigned a grammatical relation (e.g., a noun as head of an NP). The class to be predicted is the grammatical relation holding between this phrase and the verb.

5.3.1 Relation finder architecture

An instance for such a pair of words is constructed by extracting a set of feature values from the sentence. The instance contains information about the verb and the focus: a feature for the word form and a feature for the POS of both. It also has similar features for the local context of the focus. Experiments on the training data suggest an optimal context width of two words to the left and one to the right, as was the case for chunking. In addition to the lexical and the local context information, superficial information about clause structure was included as well: the distance from the verb to the focus, counted in numbers of words. A negative distance means that the focus is to the left of the verb. Other features contain the number of other verbs between the verb and the focus, and the number of intervening commas. These features were chosen by manual "feature

Struct.			Verb		Context -2			Context -1			Focus					Context +1			Class
					word	pos	cat	word	pos	cat	pr	word	pos	cat	adv	word	pos	cat	
1	2	3	4	5	6	7	8	9	10	11	12	13	14	15	16	17	18	19	
-5	0	2	org.	VBD	-	-	-	-	-	-	-	surp.	RB	ADVP	-	,	,	-	-
-3	0	1	org.	VBD	surp.	RB	ADVP	,	,	-	-	Miller	NNP	NP	-	,	,	-	-
-1	0	0	org.	VBD	Miller	NNP	NP	,	,	-	-	who	WP	NP	-	org.	VBD	VP	NP-SBJ
1	0	0	org.	VBD	who	WP	NP	org.	VBD	VP	-	conf.	NN	NP	-	York	NNP	PP	NP
2	0	0	org.	VBD	org.	VBD	VP	conf.	NN	NP	IN	York	NNP	PP	LOC	,	,	-	-

Table 5.4: The first five instances for the example sentence. Features 1–3 are the features for distance and intervening VPs and commas. Features 4 and 5 show the verb and its POS. Features 6–8, 9–11 and 17–19 describe the context words/chunks, features 12–16 the focus chunk. Empty contexts are indicated by the "-" for all features. Some words are abbreviated.

engineering" (Buchholz, 2002). Table 5.4 shows some of the instances corresponding to the following sentence (POS tags after the slash, chunks denoted with square and curly brackets, and adverbial functions after the dash). All this information is provided by tagging and chunking:

> [ADVP Not/RB surprisingly/RB ADVP] ,/, [NP Peter/NNP Miller/NNP NP] ,/, [NP who/WP NP] [VP organized/VBD VP] [NP the/DT conference/NN NP] {PP-LOC [Prep in/IN Prep] [NP New/NNP York/NNP NP] PP-LOC} ,/, [VP does/VBZ not/RB want/VB to/TO come/VB VP] {PP-DIR [Prep to/IN Prep] [NP Paris/NNP NP] PP-DIR} [Prep without/IN Prep] [VP bringing/VBG VP] [NP his/PRP$ wife/NN NP].

5.3.2 Results

Table 5.5 shows the results of the experiments. In the first row, only POS tag features are used. Other rows show the results when adding several types of chunk information as extra features. The more structure is added, the better the results: precision increases from 60.7% to 74.8%, recall from 41.3% to 67.9% — in spite of the fact that the added information is not always correct, because it was predicted by other modules of the shallow parser. With "perfect" information from these modules, a precision of 86.3% and recall of 80.8% would be attainable.

In Buchholz (2002), the MBLP approach to GR finding described here was further investigated and optimized both for accuracy and for efficiency

5.4. CONCLUSION

Structure in input	Feat.	Δ	Prec.	All Rec.	F	Subj. F	Obj. F	Loc. F	Temp. F
words and POS only	13	6.1	60.7	41.3	49.1	52.8	49.4	34.0	38.4
+VP chunks	17	6.6	63.8	47.9	54.7	62.9	51.5	39.0	42.8
+NP chunks	17	4.2	65.9	55.7	60.4	64.1	75.6	37.9	42.1
+VP chunks	17	4.5	72.1	62.9	67.2	78.6	75.6	40.8	46.8
+ADVP/ADJP chunks	17	4.4	72.1	63.0	67.3	78.8	75.8	40.4	46.5
+Prep chunks	17	4.4	72.5	64.3	68.2	81.2	75.7	40.4	47.1
+PP chunks	18	3.6	73.6	65.6	69.3	81.6	80.3	40.6	48.3
+ADVFUNCs	19	3.6	74.8	67.9	71.2	81.8	81.0	46.9	63.3

Table 5.5: Results (in %) of grammatical relation assignment with increasing levels of structure in the test data added by earlier modules in the cascade. Columns show the number of features in the instances, the average distance between the verb and the focus element, precision, recall and F-score (with $\beta = 1$) over all relations, and F-score over four selected relations. Reproduced from (Buchholz et al., 1999).

by careful feature engineering, system design adaptation, and algorithm parameter optimization, increasing precision and recall to 80% and 86.5%, respectively. Finally, in a surprising learning curve result, Van den Bosch and Buchholz (2002) show that when sufficient training data is available, words only can be used successfully as features to predict constituent structure and grammatical relations, obviating the need for POS tags.

5.4 Conclusion

From the point of view of text mining applications, robust shallow parsing seems at present to yield more useful results than deep parsing, by providing for each sentence what the main constituents and the grammatical relations between them are. In this chapter, we showed that an MBLP approach to shallow parsing is feasible by dividing the problem into a number of subproblems (tagging, chunking, and relation finding), each of which can be handled by a memory-based classifier.

This combination of memory-based classifiers, especially when provided in a server-client set-up, can be extended with domain-specific tokenizers and named-entity recognizers to provide a flexible shallow understanding architecture for use in text mining applications. In the current version of our memory-based shallow parser, the first paragraph of this book receives the following analysis:

[NP-SBJ-1 *This*/DT *book*/NN NP-SBJ-1] [VP-1 *presents*/VBZ VP-1] [NP-OBJ-1 *a*/DT *simple*/JJ *and*/CC *efficient*/JJ *approach*/NN NP-OBJ-1] [PP *to*/TO PP] [VP-2 *solving*/VBG VP-2] [NP-OBJ-2 *Natural*/NNP *Language*/NNP *Processing*/NNP *problems*/NNS NP-OBJ-2] ./.

[NP-SBJ-1 *The*/DT *approach*/NN NP-SBJ-1] [VP-1 *is*/VBZ *based*/VBN VP-1] {PNP [PP *on*/IN PP] [NP *the*/DT *combination*/NN NP] PNP} {PNP [PP *of*/IN PP] [NP *two*/CD *powerful*/JJ *techniques*/NNS NP] PNP} :/: [NP *the*/DT *efficient*/JJ *storage*/NN NP] {PNP [PP *of*/IN PP] [NP *solved*/VBN *examples*/NNS NP] PNP} {PNP [PP *of*/IN PP] [NP *the*/DT *problem*/NN NP] PNP} ,/, *and*/CC [NP *similarity-based*/JJ *reasoning*/NN NP] {PNP [PP *on*/IN PP] [NP *the*/DT *basis*/NN NP] PNP} {PNP [PP *of*/IN PP] [NP *these*/DT *stored*/VBN *examples*/NNS NP] PNP} [VP-2 *to*/TO *solve*/VB VP-2] [NP-OBJ-2 *new*/JJ *ones*/NNS NP-OBJ-2] ./.

This chapter finishes our selection of illustrations of the MBLP approach to NLP tasks, started in the previous chapter. Our main goal was to provide a few salient examples showing how to make NLP problems fit the memory-based approach. In the next chapter, we return to a machine learning perspective and discuss the eager-lazy learning dimension. We show empirically that highest accuracy can be achieved in a lazy learning approach like MBLP.

5.5 Further reading

A lot of work since Ramshaw and Marcus (1995) has focused on machine learning approaches to shallow parsing. A good place to start is the papers and references in the special issue of the *Journal of Machine Learning Research* on this topic (Hammerton et al., 2002). In the context of the CoNLL shared tasks, training and test data for chunking and clause boundary detection is available, and many machine learning results on these data can be accessed through the SIGNLL web sit [1].

The memory-based shallow parsing approach described in this chapter has been used successfully in question answering (Buchholz & Daelemans, 2001), and is currently being adapted and applied in projects on information extraction from biomedical text, automatic subtitling by summarization (Daelemans et al., 2004a), ontology extraction from text (Reinberger

---

[1]SIGNLL is the Association for Computational Linguistics' special interest group on machine learning of language, http://www.aclweb.org/signll/

et al., 2004), spoken language parsing (Canisius & Van den Bosch, 2004), and other applications. An area of current research is to integrate an MBLP module for PP-attachment as an additional component in the shallow parser. Earlier work has shown that this task in isolation is indeed feasible in a memory-based approach (Zavrel & Daelemans, 1997; Van Herwijnen et al., 2004; Kokkinakis, 2000) and could be integrated into the current architecture in a way similar to the integration of grammatical relation finding.

An alternative approach to memory-based chunking and relation finding has been proposed in Argamon et al. (1999) under the name of memory-based sequence learning. The method is based on a search among possible bracketings of sentences, keeping all training data available. The approach can be seen as a linear algorithmic simplification of the DOP memory-based approach to full parsing discussed in chapter 2.

Moving from shallow parsing to full parsing by extending the memory-based chunking approach iteratively to approximate full parsing has not been entirely successful yet (see for example Tjong Kim Sang, 2002 for an empirical investigation). In contrast, the OCTOPUS parser for Chinese (Streiter, 2001b; Streiter, 2001a) uses complete parse trees as memory instances, and retrieves nearest neighbors by matching sequences of keywords, where processes of alignment, nearest neighbor adaptation, and chunking cooperate in providing a parse tree for the input sentences. Another approach to memory-based learning with complete parse trees as "classes" is TüSBL (Kübler, 2004). In this system conventional tagging and chunking are used to provide features for a new similarity metric on a dynamically computed set of features working on complete parse trees as examples.

An alternative way to full parsing defined as a classification-based approach is the framework of shift-reduce parsing, where the next parser decision is predicted given the current state of the parse and the local context as instances. The different steps of the derivations of a parse are used as training instances. A memory-based approach was explored in Veenstra and Daelemans (2000) and recently developed in the context of Swedish and English dependency parsing with the MALT parser (Nivre et al., 2004; Nivre & Scholz, 2004). In combination with alternative parsing methods, memory-based learning has also been found useful for tasks such as the enrichment of parser output (Jijkoun & de Rijke, 2004).

Chapter 6

Abstraction and generalization

The concepts of abstraction and generalization are tightly coupled to Ockham's razor, a medieval scientific principle, which is still regarded in many branches of modern science as fundamentally true. Sources quote the principle as "non preterio necessitate delendam", or freely translated in the imperative form, *delete all elements in a theory that are not necessary*. The goal of its application is to maximize economy and generality: it favors small theories over large ones, when they have the same expressive power. The latter can be read as 'having the same generalization accuracy', which, as we have exemplified in the previous chapters, can be estimated through validation tests with held-out material.

A twentieth-century incarnation of Ockham's razor is the minimal description length (MDL) principle (Rissanen, 1983), coined in the context of computational learning theory. It has been used as the leading principle in the design of decision tree induction algorithms such as C4.5 (Quinlan, 1993) and rule induction algorithms such as RIPPER (Cohen, 1995). The goal of these algorithms is to find a compact representation of the classification information in the given learning material that at the same time generalizes well to unseen material. C4.5 uses decision trees; RIPPER uses ordered lists of rules to meet that end.

In contrast, memory-based learning is not minimal – its description length is equal to the amount of memory it takes to store the learning examples. Keeping all learning examples in memory is all but economical. Classification in memory-based learning, on the other hand, does constitute a certain type of abstraction from the full data; more specifically, abstraction in memory-based classification is of a local, impermanent nature. With each classification of a new instance, the total memory available is reduced to a small subset of k-nearest neighbors, thereby forgetting (for a moment) the rest of the material. The impermanence of this type of selective forgetting is usually not considered abstraction, which has a more permanent connotation. However, permanency is arguably not a defining criterion of abstraction.

The first section of this chapter, section 6.1, is devoted to an exemplifying comparison of the impermanent, non-economical abstraction in standard memory-based learning versus the permanent, Economical abstraction in rule induction. In the section we compare standard memory-based learning against IGTREE, the IB1 approximation introduced in chapter 4, in which parts of the memory are compressed in a lossy decision tree structure (lossy in the sense that the example base cannot be reconstructed from it), and against RIPPER (Cohen, 1995), an efficient rule learner. In both comparative experiments we compare generalization accuracies (or other derived performance measures, where appropriate) on a range of language learning tasks, and run additional analyses that show that memory-based learning performs equally well or better than its permanently-abstracting counterparts.

In section 6.2 we then introduce editing, a k-NN-internal method to reduce the amount of memory needed by removing particular examples from memory. We describe a method to estimate the likelihood that certain examples will be good or bad neighbors in classification, on the basis of which the bad examples could be removed. Through experiments on the six benchmark tasks we show that editing examples eventually leads to performance loss, but that some editing is possible without harm. Also, we show that the most important examples to keep in memory are the ones that have many nearest neighbors of their own class, but also several nearest neighbors of different classes – these are the same type of examples that are stored as support vectors in support vector machines (Cristiani & Shawe-Taylor, 2000).

To conclude the chapter, in section 6.4 we explore a method to create generalized examples, a less destructive permanent abstraction method in memory-based learning than editing. Generalized examples are generated from sets of nearest-neighbor same-class examples, and replace the individual examples they are made of. Only the co-occurrence information of feature values in individual examples is forgotten, while at the same time storage costs are considerably reduced. This algorithmic variant is shown to be generally harmless, yielding impressive compression rates. Also, we find that the approach is in effect a pre-compiler for k; it is possible to leave k at 1, and perform the k-NN classification rule on the generalized examples, which each represent a locally appropriate compilation of variable amounts of same-class nearest neighbors.

6.1 Lazy versus eager learning

Rule induction is the umbrella term for a class of supervised machine learning algorithms that adhere quite faithfully to the MDL principle. Together with algorithms for the top-down induction of decision trees, rule induction algorithms are often referred to as "eager" learners, which invest considerable effort in building minimally-sized models, which at the same time are estimated to generalize well to unseen data. The resulting classifier, a decision tree or a set of ordered rules, can be quite readable from a human perspective. This attractive feature is in stark contrast with the lack of generalizable information in the internal models of IB1, or for that matter in the models learned by support vector machines, linear threshold classifiers, or maximum-entropy models, which all harbor internal models composed of matrices or networks of real numbers of which the meaning is quite hermetic.

The transparency and readability of models produced by decision tree induction or rule induction is mirrored in the simplicity of both of their classification procedures. Classification in rule-induction classifiers is typically based on the firing of a rule on a test instance, triggered by matching feature values at the left-hand side of the rule. From a human perspective this is typically easy to understand; to read and interpret a rule is arguably easier than to inspect which k-nearest neighbors have determined IB1's classification of an individual test instance. It is also easier than understanding the outcome of a linear threshold function, or multiplications of estimated probabilities, leading to the classifications of individual test instances.

The ease of classification in rule induction is counterweighted by a complex learning algorithm. This is entirely in contrast with MBL, which has a very simple learning algorithm (store examples in memory) but a costly classification algorithm. In rule induction, the learning problem is to search for an optimal set of rules. Rules can be of various normal forms, but even the simpler forms, such as conjunctive normal form (CNF), open up a vast space of possible rules exponential in the number of feature values. Rules are also typically ordered (in classification, the first rule in the ordering that matches the test instance determines the classification). The appropriate content and ordering of rules is typically very hard to

6.1. LAZY VERSUS EAGER LEARNING

find. Therefore, at the heart of most rule induction systems, strong heuristic search algorithms are employed to attempt to minimize search through the space of possible rule sets and orderings.

As an example rule induction algorithm, we run comparative experiments with RIPPER (Cohen, 1995). RIPPER belongs to the family of sequential covering rule inducers. It induces subsets of rules per class, in a predetermined class ordering. By default, the ordering is from low-frequency classes to high-frequency classes, leaving the most frequent class as the default rule (which is generally beneficial for the total description length of the rule set). Within a class, RIPPER searches for the shortest rules with the best coverage and accuracy. Central to RIPPER's heuristic algorithm is splitting the training set in two. On the basis of one part it induces a list of candidate rules. When a candidate rule classifies instances in the other part of the split training set below a threshold, and/or it is too long to be estimated to be useful, it is discarded. The search for rules in RIPPER is guided heuristically by information gain measurements of certain features.

6.1.1 Benchmark language learning tasks

To produce a sensible and broad pool of comparative data, we apply both IB1 and RIPPER to six NLP tasks ranging from morpho-phonological tasks to semanto-syntactic tasks, varying in scope (word level and sentence level) and basic type of example encoding (non-windowing and windowing). We briefly describe the six tasks here and provide some basic data set specifications in Table 6.1. Note that some of these data sets have been used for illustrative purposes and for experiments described in earlier chapters; in the subsequent chapter 7 they are used again in several series of experiments.

1. GPLURAL, the formation of the plural form of German nouns, was introduced in section 3.1. The task is to classify a noun (represented in a single example, of which the seven features encode the contents of its two final syllables and its gender) as mapping to one out of eight classes, representing the noun's plural formation. We use the same 50%–50% split in 12,584 training examples and 12,584 test instances as used in section 3.1. Generalization performance is measured in accuracy, viz. the percentage of correctly classified test instances.

2. DIMIN, Dutch diminutive formation, uses a similar scheme to the one used in the GPLURAL task to represent a word as a single example.

The task and data were introduced in (Daelemans et al., 1997a). A noun, or more specifically its phonemic transcription, is represented by its last three syllables, which are each represented by four features: (1) whether the syllable is stressed (binary), (2) the onset, (3) the nucleus, and (4) the coda. The class label represents the identity of the diminutive inflection, which is one out of five (-je, -tje, -etje, -pje, or -kje). For example, the diminutive form of the Dutch noun beker (cup) is bekertje (little cup). Its phonemic representation is ['bekər]. The resulting example is _ _ _ _ + b e _ − k ə r tje. The data are extracted from the CELEX-2 lexical database (Baayen et al., 1993). The training set contains 2,999 labeled examples of nouns; the test set contains 950 instances. Again, generalization performance is measured in accuracy, viz. the percentage of correctly classified test instances.

3. MORPH was introduced earlier in section 4.2. It represents the morphological analysis of Dutch words; i.e., to analyze abnormaliteiten as

$$[\text{abnormaal}]_A \ [\text{iteit}]_{A\_\to N} \ [\text{en}]_{plural}$$

The task combines segmentation, part-of-speech tagging of morphemes, and undoing spelling changes to recover the stem morphemes. Generalization performance is measured in the F-score on correctly identified (segmented, labeled, orthographically reconstructed) morphemes in unseen test words.

4. PP, prepositional-phrase attachment, was introduced in section 1.2 and is the classical benchmark data set introduced in (Ratnaparkhi et al., 1994). The data set is derived from the Wall Street Journal Penn Treebank (Marcus et al., 1993). All sentences containing the pattern "VP NP PP" with a single NP in the PP were converted to four-feature examples, where each feature contains the head word of one of the four constituents, yielding a "V N1 P N2" pattern such as *"each pizza with Eleni"*, or *"eat pizza with pineapple"*. Each example is labeled by a class denoting whether the PP is attached to the verb or to the N1 noun in the treebank parse. We use the original training set of 20,800 examples, and the test set of 3,097 instances. Noun attachment occurs slightly more frequently than verb attachment; 52% of the training examples and 59% of the test examples are noun attachment cases. Generalization performance is measured in terms of accuracy (the percentage of correctly classified test instances).

6.1. LAZY VERSUS EAGER LEARNING

5. CHUNK, introduced earlier in section 5.2, , is the task of splitting sentences into non-overlapping syntactic phrases or constituents, e.g., to analyze the sentence "He reckons the current account deficit will narrow to only $ 1.8 billion in September." as

   [He]$_{NP}$ [reckons]$_{VP}$ [the current account deficit]$_{NP}$ [will narrow]$_{VP}$ [to]$_{PP}$ [only $ 1.8 billion]$_{NP}$ [in]$_{PP}$ [September]$_{NP}$.

   The data set, extracted from the WSJ Penn Treebank through a flattened, intermediary representation of the trees (Tjong Kim Sang & Buchholz, 2000), contains 211,727 training examples and 47,377 test instances. The examples represent seven-word windows of words and their respective part-of-speech tags computed by the Brill tagger (Brill, 1992) (which is trained on a disjoint part of the WSJ Penn Treebank), and each example is labeled with a class using the IOB type of segmentation coding as introduced in (Ramshaw & Marcus, 1995) and introduced earlier in section 5.2. Generalization performance is measured by the F-score on correctly identified and labeled constituents in test data, using the evaluation method originally used in the "shared task" sub-event of the CoNLL-2000 conference (Tjong Kim Sang & Buchholz, 2000) in which this particular training and test set were used.

6. NER, named-entity recognition, is to recognize and type named entities in text. With NER, the sentence "U.N. official Ekeus heads for Baghdad." is analyzed as

   [U.N.]$_{organization}$ official [Ekeus]$_{person}$ heads for [Baghdad]$_{location}$.

   We employ the English NER shared task data set used in the CoNLL-2003 conference, again using the same evaluation method as originally used in the shared task (Tjong Kim Sang & De Meulder, 2003). This data set discriminates four name types: persons, organizations, locations and "miscellany names", capturing all other named entities under this one label. The data set is a collection of newswire articles from the Reuters Corpus, RCV1[1]. The given training set contains 203,621 examples; as test set we use the "testb" evaluation set which contains 46,435 examples. Examples represent seven-word windows of words with their respective predicted part-of-speech tags (no other

---

[1] Reuters Corpus, Volume 1, English language, 1996-08-20 to 1997-08-19.

Task	Number of Examples	Number of Features	Range of number of values	Number of classes
GPLURAL	12,584	7	8 – 81	8
DIMIN	2,999	12	2 – 69	5
MORPH	2,888,255	7	49 – 55	3,831
PP	20,801	4	66 – 5,451	2
CHUNK	211,727	14	44 – 19,122	22
NER	203,621	14	45 – 23,623	8

Table 6.1: Properties of the training sets representing the six learning tasks: numbers of examples, features, minimum and maximum number of values over all features, and classes.

task-specific features such as capitalization identifiers or seed list features were used). Class labels use the IOB segmentation coding coupled with the four possible name type labels. Again analogous to the CHUNK task, generalization performance is measured by the F-score on correctly identified and labeled named entities in test data.

Table 6.1 illustrates that apart from covering different levels of NLP, the six tasks offer a wide variety of data properties with differences of three to five orders of magnitude in numbers of instances, values, and classes. The smallest data set, DIMIN, is about a thousand times smaller than the biggest, MORPH. The numbers of values of features range from two (the presence marker of word stress on the final syllable in the DIMIN task) to several thousands (all features representing words in PP, CHUNK, and NER). The number of classes also ranges from two (PP: noun or verb attachment) to almost four thousand (the many complex classes of the MORPH task). Only the number of features lies between four and 14, reflecting the common hypothesis across the tasks that they are sufficiently learnable by a fairly small set of features. All tasks are represented by small contexts of surface features (letters and words), possibly augmented by somewhat less shallow features (the POS tags with CHUNK and NER). This is a concededly simple hypothesis that we maintain essentially for simplicity; we also do not include any special non-surface features (e.g., seed list features in NER). However, it should be kept in mind that nominal features with thousands of values in MBLP like for the PP, CHUNK, and NER data sets, would correspond to thousands of different features in binary learners like maximum entropy learning and support vector machines.

6.1.2 Forgetting by rule induction is harmful in language learning

One way to study the influence of ignoring exceptional instances on generalization accuracy is to compare IB1 to an inductive algorithm that abstracts from certain instances while building its model. We already introduced RIPPER
as our example rule induction algorithm which we are going to use in this comparison. In addition, we also compare with IGTREE, introduced in section 4.1.2 – an approximation of k-NN classification in IB1, which abstracts from certain individual instances in building its tree. We compare the generalization performance of the three algorithms using the six data sets described in the previous subsection. In this subsection we discuss the results of this comparison, and the influence of a parameter of RIPPER that directly affects the minimal number of examples an induced rule is allowed to cover, on generalization accuracy.

We compare the performances of IB1, IGTREE, and RIPPER on the six benchmark data sets first by running single tests in which the algorithmic parameters of the three algorithms are automatically optimized in a model selection procedure described later in the book, in section 7.1. Note that all experiments are performed on single train–test splits, and that for three data sets (GPLURAL, DIMIN, PP) the reported performance metric is generalization accuracy (the percentage of correctly classified test instances), while for the other three data sets (MORPH, CHUNK, NER) the generalization performance is measured by an F-score averaging the precision and recall of the relevant structure of the task at hand (e.g., labeled, segmented stem morphemes with MORPH; chunked and labeled base phrases in CHUNK; chunked and labeled named entities with NER).

Table 6.2 displays the generalization accuracies, measured in percentages of correctly classified test instances, for IB1, IGTREE, and RIPPER on the six tasks. We make the following observations:

1. IB1 outperforms IGTREE on all tasks except GPLURAL; their scores on MORPH are fairly close.

2. IB1 outperforms RIPPER on all tasks.

3. IGTREE outperforms RIPPER on GPLURAL, MORPH, and NER. Their performances are close on DIMIN and PP, and RIPPER outperforms IGTREE on the CHUNK task.

Clearly, the two abstracting algorithms IGTREE and RIPPER do not outperform the non-abstracting algorithm IB1 on any task. It is less clear,

	Performance	Generalization performance (%)		
Task	metric	IB1	IGTREE	RIPPER
GPLURAL	accuracy	94.0	94.3	91.0
DIMIN	accuracy	97.6	96.6	96.7
MORPH	F-score	70.1	69.9	38.4
PP	accuracy	80.7	76.7	76.1
CHUNK	F-score	91.9	87.6	89.5
NER	F-score	77.2	66.6	55.5

Table 6.2: Generalization performances (accuracies or F-scores) on the six benchmark tasks, by IB1 (with gain-ratio feature weighting, MVDM, and $k = 1$), IGTREE (with gain-ratio feature ordering), and RIPPER (with default settings).

except in the case of the MORPH task, whether RIPPER's abstraction by rules is more harmful than IGTREE's decision tree abstraction. The two algorithms differ quite significantly in the amount of abstraction they attain, as the memory usage of the three algorithms underlines. Table 6.3 displays four memory statistics that are reasonably comparable. First, the table displays for all six benchmark tasks the multiplication of the number of examples times the number of features (plus one, for the class label), representing the "worst-case" upper bound costs of storing all examples in a flat structure in memory. Second, the table displays the actual number of nodes in the internal IB1 tree used for searching for the nearest neighbors. The memory compression factors of this internal storage range between 24% (NER) to 92% (MORPH), providing rough indications of the internal redundancy of the training sets. The third statistic in Table 6.3 lists the numbers of nodes in the IGTREE trees – pruned versions of the IB1 trees. IGTREE compression factors are large, compared to the upper-bound storage; between 95% (PP) and 99.7% (NER). Fourthly, RIPPER attains even larger compression rates, under the hundredth percent as compared to the upper bound, but also between 66% and 99.6% as compared to IGTREE. The RIPPER counts in Table 6.3 are counts of the number of conditions, rather than numbers of rules, because a condition (in RIPPER's default grammar, a test on a feature value) can be likened to a node in a tree.

In view of these results it is easy to be amazed at the results of IGTREE and RIPPER, and especially of the latter — consider, for example, its reasonable performance on the CHUNK task. With just 721 conditions in

6.1. LAZY VERSUS EAGER LEARNING

Task	Number of examples × feat.	Number of tree nodes IB1	IGTREE	Number of RIPPER cond.
GPLURAL	100,672	27,759	1,311	322
DIMIN	38,987	22,705	191	64
MORPH	23,106,040	1,771,841	107,694	11,491
PP	104,005	68,648	5,061	18
CHUNK	3,175,905	2,316,308	20,090	721
NER	3,054,315	2,325,286	9,178	443

Table 6.3: Memory usage statistics of IB1, IGTREE, and RIPPER applied to the six benchmark tasks: the number of examples times features (raw storage in IB1), the number of nodes in the IB1 tree and the pruned IGTREE tree, and the number of RIPPER conditions.

416 rules, it attains a higher F-score than IGTREE, which uses over five thousand nodes in its tree. Moreover, it is just 2 points of F-score short of IB1's, which needs an internal tree with over two million nodes. Yet, neither of the two outperforms IB1; RIPPER's amazing compression rates have a price.

Increased abstraction in IB1 and RIPPER

The general comparative results of the previous subsection do not offer a complete picture. For one, RIPPER offers a parameter that enforces more or less abstraction, namely by forcing rules to cover at least a certain threshold number of examples. By default, as also used in the experiments above, this parameter is set to 1, i.e., RIPPER is allowed to induce rules that cover only one example. However, RIPPER will typically avoid inducing example-specific rules, since these extremely low-coverage rules are very inefficient in terms of storage costs; after all, RIPPER's guiding principle is minimal description length.

Still, RIPPER can abstract even more than it does by default. We performed additional experiments with IB1 and RIPPER in which we enforce in both algorithms that classifications are based on more than a single instance. RIPPER has a parameter that sets the minimal number of examples that a rule is allowed to cover (henceforth named F, after RIPPER's command-line parameter switch). Changing this threshold from its default of 1 to, for example, 2, causes RIPPER to abandon certain rules that covered single examples. Abandoning these example-specific rules effectively means forgetting about the existence of the now uncovered single examples. In IB1, classifications can be enforced to be based on more than a single instances by setting $k > 1$. No examples are actually abstracted from with $k > 1$: examples only play relatively smaller roles in determining the classification.

We increased both IB1's k parameter and RIPPER's F parameter in the following numerical series: 1, 2, ..., 9, 10, 15, ..., 45, 50, 60, ..., 90, 100, 150, ..., 450, 500. As in the previous subsection, RIPPER was run with default parameter settings, and IB1 was run with the MVDM similarity metric and gain-ratio feature weighting. Figure 6.1 displays the bundle of experimental outcomes of RIPPER and IB1 on the six benchmark data sets.

Our previous observation that RIPPER never outperforms IB1 is mirrored by the results with increasing thresholds on the minimal number of instances a rule must cover in RIPPER. With most tasks the difference between IB1 and RIPPER is aggravated when this threshold is increased. Interestingly, the effect of a higher k on IB1's performance is mixed; on the two (morphological) tasks GPLURAL and MORPH, increasing k has a clear detrimental effect on performance. On the other hand, IB1's score is hardly affected by the particular setting of k for the other tasks, and even minor improvements can be seen with all four of them; performances appear to peak roughly at around $k = 10$.

In sum, these results suggest that abstraction by rules is more harmful than memorizing all examples; moreover, the detrimental effect of abstraction tends to increase along with RIPPER's adaptable bias to force rules to cover more examples, i.e., to be more general. At the same time, in some cases increasing k does occasionally lead to losses in generalization performance. For the tasks in which this is the case, GPLURAL and MORPH, it is apparently harmful to base classifications on more than just one nearest neighbor. In the data representing these two tasks certain useful single examples exist that apparently have a deviating class from their direct neighborhood; they can only exert their positive influence as a nearest neighbor when $k = 1$, i.e., when their different-class neighbors cannot outweigh them in voting for the class. In these cases, $k > 1$ has a negative effect similar to RIPPER forcing rules to cover more than a single example.

6.2. EDITING EXAMPLES

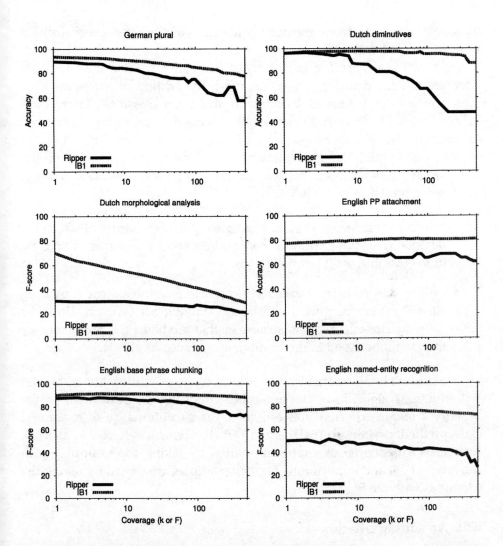

Figure 6.1: Generalization accuracies (in terms of % of correctly classified test instances) and F-scores, where appropriate, of RIPPER with increasing F parameter and IB1 with increasing k parameter, on the six benchmark data sets. The x-axis, representing the F and k parameters, uses a logarithmic scale.

6.2 Editing examples

Although the previous experiments suggest that keeping full memory is a good idea, they do not rule out that certain examples can in fact be safely removed from memory without affecting the generalization performance of IB1. The idea of removing, or editing, examples from memory without harming performance or even for noise reduction has been around for decades. In earlier work the editing of examples from memory in memory-based learning or the k-NN classifier (Hart, 1968; Wilson, 1972; Devijver & Kittler, 1980) is argued to serve two objectives: to minimize the number of examples in memory for reasons of speed or storage, and to minimize generalization error by removing noisy examples, prone to being responsible for generalization errors. Two basic types of editing can be found in the literature:

- **Editing superfluous regular examples**: delete examples of which the deletion does not harm the classification accuracy of their own class in the training set (Hart, 1968).

- **Editing unproductive exceptions**: delete examples incorrectly classified by their neighborhood in the training set (Wilson, 1972), or roughly vice-versa, delete examples that are bad class predictors for their neighborhood in the training set (Aha et al., 1991).

We present experiments in which both types of editing are employed within the IB1 algorithm. The two types of editing are performed on the basis of a single criterion that estimates the exceptionality of examples: class prediction strength (Salzberg, 1990) (henceforth referred to as CPS). Estimated exceptional examples are edited by taking the examples with the lowest CPS, and superfluous regular examples are edited by taking the examples with the highest CPS.

CPS: An editing criterion

CPS is an estimate of the (degree of) exceptionality of examples. It is closely tied to the k-NN classification rule; it makes used of it. CPS offers an estimate of how well an example predicts the class of all other examples within the training set. In earlier work, CPS has been used as a criterion for removing examples in memory-based learning algorithms, e.g., in IB3 (Aha et al., 1991); or for weighting examples in the memory-based EACH algorithm (Salzberg, 1990). We adopt the class-prediction strength estimate as proposed by Salzberg (1990), who defines it as the ratio of the number of times the example is a nearest neighbor of another example with the same class and the number of times that the example is the nearest neighbor

6.2. EDITING EXAMPLES

of another example regardless of the class. An example type with class-prediction strength 1.0 is a perfect predictor of its own class; a class-prediction strength of 0.0 indicates that the example type is a bad predictor of classes of other examples, presumably indicating that the example type is exceptional. We perform a simple Laplace correction to account for the intuitive fact that a 1.0 CPS value originating from a "100 out of 100" performance score should count higher than a "1 out of 1" performance score; we compute CPS as in Equation 6.1:

$$cps(X) = \frac{correct\_NN(X) + 0.5}{NN(X) + 1.0} \quad (6.1)$$

where $correct\_NN(X)$ is the number of times example X was the nearest neighbor of another example in the training set having the same class, and NN is the number of times example X was the nearest neighbor of another example in the training set regardless of the class. These numbers were counted in a leave-one-out experiment (cf. section 3.5) on the training set.

One might argue that bad class predictors can be edited safely from the example base. Likewise, one could also argue that examples with a maximal CPS could be edited to some degree too without harming generalization: strong class predictors may be abundant and some may be safely forgotten since other example types near to the edited one may continue and replace the class predictions of the edited instance type. Figure 6.2 offers a visualization of the effects of both types of editing. Removing all examples that have a different-class nearest neighbor (displayed in the lower left of Figure 6.2) results in the area between the two classes being wiped clean, leaving a smoother effective k-NN class border in the middle. In contrast, removing all examples of which all nearest neighbors bordering its own locally spanned area (its Voronoi cell, or Dirichlet tile, Okabe et al., 2000) are of the same class, results in the inner areas of the black and white class spaces being wiped clean, while the original class border remains identical (as displayed in the lower right of Figure 6.2).

To test the utility of CPS as a criterion for justifying the forgetting of specific training examples, we performed a series of experiments in which IB1 is applied to three of our six data sets: GPLURAL, DIMIN, and PP. With each data set we set IB1's parameters at different settings which were estimated in a validation experiment to be optimal – we refer again to section 7.1 for a detailed description of this validation method. With all three tasks, the MVDM similarity function was used. Inverse-linear distance weighting was used with GPLURAL and DIMIN; inverse distance weighting

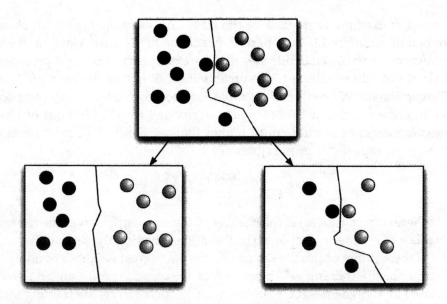

Figure 6.2: Visualization of two types of editing starting from a given memory with black and white class labeled examples (top). Left: after editing all examples with a different-class nearest neighbor. Right: after editing all examples with a same-class neighbor.

was used with PP. For GPLURAL, k was set to 5; with DIMIN, $k = 11$, and with PP, $k = 7$. The CPS statistics on good neighborship are based on these settings of k.

In the editing experiments, each of the three tasks' training sets is systematically and incrementally edited using the ranked CPS metrics obtained for all training examples from the two sides of the ranking: in one experiment, examples are deleted from the ones with the lowest to the ones with the highest CPS, while in the other experiment examples are deleted starting with the ones with the highest, and ending with the ones with the lowest CPS. We edit at all occurring thresholds of CPS, starting from the set of examples that has an above-zero CPS value. A certain percentage of examples in each of the three data sets does not actually receive a CPS value since they are never used as a nearest neighbor in the leave-one-out experiment on the training data; these examples (up to about 40% of all examples in the case of the GPLURAL task) are edited at once. With each subsequent edit, one experiment is performed in which IB1 is trained on the edited training set, and tested on the normal, constant test set. As a control,

6.2. EDITING EXAMPLES

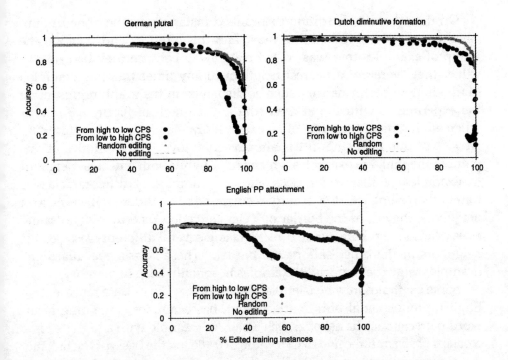

Figure 6.3: Generalization accuracies by IB1 on GPLURAL, DIMIN, and PP with incremental percentages of edited example tokens, according to the CPS editing criterion, from high to low CPS and vice versa, and using random incremental editing.

we also generate an incremental editing sequence in which examples are randomly deleted one by one. The results of these series of experiments are displayed in Figure 6.3.

The general trends we observe in the results are that editing on the basis of CPS from either side of the rank never boosts generalization accuracy, and that editing is ultimately harmful to generalization accuracy at levels of about 40% and upwards. Generalization accuracies tend to decrease more at higher editing rates. Another rather surprising observation in all three cases is that editing examples randomly is generally less harmful than editing from either side of the CPS rank at the highest editing rates. Editing up to 80% of PP examples randomly, or even 90% of the DIMIN examples, does not result in a lower score than obtained with training on all examples. At the same editing levels, editing examples from either side of the CPS rank does lead to lower accuracies on test data.

On the other hand, editing CPS-ranked instances can be done unpunished up to rather high editing rates, and this effect extends beyond the group of examples that was edited right away because they had no CPS value (they were not a nearest neighbor to any other training example). Markedly low performances, such as displayed in the graph representing the experiments with the PP task (the lower graph of Figure 6.3), occur when editing more than 40% of all examples ranked from high to low CPS. At 40% of this rank, the CPS is already quite low. The examples at this point in the rank are examples which are, roughly, both a nearest neighbor to examples of their own class and to examples of different classes. Rather than being isolated examples surrounded by different-class nearest neighbors, they lie at the border of a class area occupied by several same-class nearest neighbors, while directly facing several different-class nearest neighbors at the other side of the border. These "class-area guarding" examples might be important examples to actually keep in memory.

Some examples may be useful as illustrations. The PP data representing English prepositional phrase attachment, has examples containing head-word patterns such as eat pizza with Eleni. An example with a high CPS is is producer of sugar. It receives a CPS value of 0.986, on the basis of the fact that it is a nearest neighbor of 177 fellow training examples, predicting a correct class (noun attachment) in 176 cases. The one nearest neighbor with a different class is accuses Motorola of turnabout, one of the few examples with the preposition of in which the prepositional phrase is actually attached to the verb.

A PP example with a very low CPS value is conducted inquiry into activities, an example of noun attachment. It is the nearest neighbor to 32 fellow training examples which all represent cases of verb attachment. Examples of verb-attachment nearest neighbors include bring charges into line, implant tissue into brain, and extend battles into towns. The conducted inquiry into activities example is a minority example of a noun-attached collocation (inquiry into) in a majority neighborhood of verb attachments with into.

To conclude our illustration, the examples displayed in Table 6.4 have a CPS value of about 0.5. Supposedly these are examples of the relatively interesting group of border training examples. All shown examples represent a specific attachment ambiguity. For instance, the first example, improve controls on branches, a case of noun attachment according to the original labeling, contains the collocationally strong improve on verb attachment pattern, but also the controls on noun attachment pattern. As the nearest-neighbor statistics in Table 6.4 show, they predict an incorrect class in about as many cases as they predict a correct class for fellow

6.2. EDITING EXAMPLES

Example	Attachment	Number of nearest neighbors same class	other class
improve controls on branches	noun	44	46
affect security for years	verb	5	5
is time for Japan	noun	48	51
owned store in Cleveland	noun	11	11
insures benefits for workers	verb	9	8

Table 6.4: Five examples with a CPS value of about 0.5, having about as many nearest neighbors with the same class as with a different class.

training examples. This does not mean, however, that the classifications these examples are involved in are of low quality or chance. In the ten times the second example affect security for years is a nearest neighbor of other training examples, in eight cases the total nearest neighbor set (with $k = 7$ for PP) votes for the correct class, despite the fact that this particular example votes for the incorrect class five times.

The editing experiments presented so far edit examples in an incremental way, according to a ranking criterion. Another view on the effects of editing can be obtained by not editing incrementally, but by holding out subgroups of examples in the same CPS rank. We performed two additional series of editing experiments holding out portions of the data. In the first experiment, we held out each consecutive 10% of the examples, ranked according to their CPS, and trained each time on the remaining 90% training data while testing on the regular test sets. In the second series, as a mirror of the first, we take only the consecutive 10% held-out sets as the training material, and tested on the regular test sets. We perform these experiments on GPLURAL, DIMIN, PP, and CHUNK; the results are displayed in Figure 6.4.

The held-out experiments show, as our earlier results, that removing most of the 10% portions does not seriously harm generalization accuracy. The mirror experiments, however, in which we use only the 10% held-out partitions as training material, do show interesting differences. For three of the four tasks the best performance is obtained with the set of examples in the third partition, between the 20% and 30% mark in the CPS rank. With the GPLURAL task, this partition contains examples with a CPS between 0.70 and 0.79. With the DIMIN task the examples range between 0.88 and 0.92; with CHUNK the range is 0.83–0.88. In these three cases, the "best" examples that as a group provide the best classifications of test data,

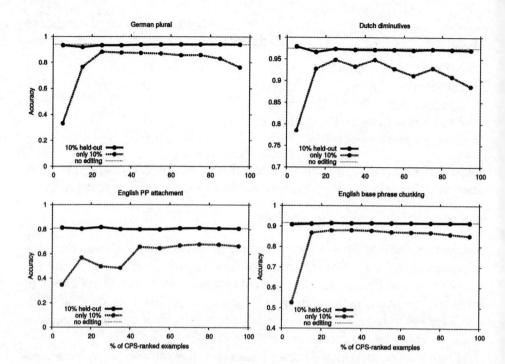

Figure 6.4: Generalization accuracies on GPLURAL, DIMIN, and PP, and F-scores on CHUNK, obtained in editing experiments in which equally-sized 10% partitions of CPS-ranked instances are edited from the full training sets (solid lines), or when these 10% partitions are used in isolation for classification (dashed lines). The thin dotted lines represent the scores of IB1 on the four tasks without editing.

represent again the important group of borderline cases with a majority of same-class friendly neighbors, but with a significant amount (between about 20% and 30%) of unfriendly nearest neighbors at the other side of the class border.

The results in Figure 6.4 also show that the examples at the far end of the CPS range are not useful as examples to solely base classifications on. This is not surprising for examples with a low CPS on the training data. Examples at the high end of the CPS scale, however, also appear to miss some generalization power. Presumably, this is because these instances are positioned at quite a large distance from the actual class borders; leaving the areas around these borders unguarded might have the effect that the actual class decisions do not follow the actual borders, leading to relatively more errors.

In sum, we observe a special importance of examples that are at the borders of their class area. The importance of these examples bears an interesting resemblance to a well-known principle of support vector machines (Cortes & Vapnik, 1995; Schölkopf et al., 1995), namely that with a proper kernel and a two-way classification task, the original set of training examples can be reduced (edited) down to a small subset of so-called support vectors, i.e., those examples that are closest (yet with a maximized margin) to the hyperplane separating the two class areas. The support vectors can be said to label and guard their class area much in the same way as the middle-CPS cases do.

6.3 Why forgetting examples can be harmful

We provide a number of analyses all contributing to an understanding of why and in what circumstances forgetting examples can be harmful to generalization accuracy. We attempt to find explanations through analyzing generic statistical properties of NLP task data, and by performing additional analyses on the outcomes of the experiments with IB1 on these data, testing whether exceptionality or typicality of examples involved in classification (estimated by CPS) relates to the quality of those classifications.

NLP tasks are hard to model and also hard to learn in terms of broad-coverage rules, because apart from obvious regularities, they also tend to contain many sub-regularities and (pockets of) exceptions. In other words, apart from a core of generalizable regularities, there is a relatively large periphery of irregularities (Daelemans, 1996). In rule-based NLP, this problem has to be solved using mechanisms such as rule ordering, subsumption, inheritance, or default reasoning (in linguistics this type of "priority to the most specific" mechanism is called the *elsewhere condition*). In the memory-based perspective this property is reflected in the high degree of disjunctivity of the example space: classes exhibit a high degree of polymorphism, or phrased alternatively, they are scattered over so many small areas in the class space that each needs at least one representative example to guard the class borders.

Degree of polymorphism

One way of estimating the degree of polymorphism, or the degree of scatteredness of classes in small class areas, is to make counts of the numbers of "friendly" (same-class) neighbors per example in a leave-one-out experiment. For each example in the six data sets a distance ranking of all other training examples is produced. Within this ranked list we note the ranking number of the first nearest neighbor with a different class. This rank number is then taken as the number of friendly neighbors surrounding the held-out example, and all friendly examples in the area are subsequently removed from the list of examples to be held out (since their counts are mutual). If, for example, a held-out example is surrounded by three examples of the same class at distance 0.0 (i.e., no mismatching feature values), followed by a fourth nearest-neighbor example of a different class at distance 0.3, the held-out example is counted as having three friendly neighbors. The counts from the six leave-one-out experiments are displayed graphically in Figure 6.5. The x-axis of Figure 6.5 denotes the numbers of friendly neighbors found surrounding examples with up to 50 friendly neighbors; the y-axis denotes the cumulative percentage of occurrences of friendly-neighbor clusters of particular sizes.

The cumulative percentage graphs in Figure 6.5 show that in all six tasks relatively many examples have but a few friendly neighbors. In the MORPH task data, over 50% of all examples have less than ten friendly nearest neighbors. In all task data sets there are more isolated examples without friendly neighbors than examples with just one friendly neighbor. The average percentage of isolated examples (i.e., the starting points of the cumulative curves) is roughly 15%. By definition these isolated examples will have a low CPS when measured on training material, if they are ever the nearest neighbor of any other example. However, this does not mean that they are by definition useless in predicting the class of new, unseen examples. Also, they are certainly useful when the same example would recur in test data – and in language data exceptions do recur.

Usefulness of exceptional examples

For our editing experiments we measured each training example's CPS. To evaluate whether CPS is actually a good estimate of the regularity or exceptionality of an example, we can measure how a training example actually performs as a nearest neighbor to test instances, rather than to

6.3. WHY FORGETTING EXAMPLES CAN BE HARMFUL

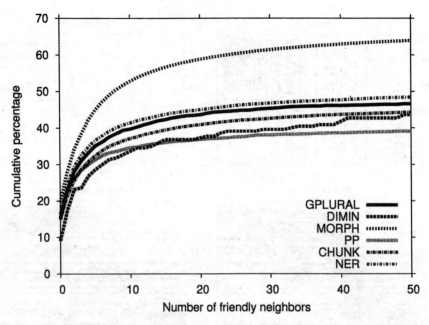

Figure 6.5: Cumulative percentages of friendly-neighbor clusters of sizes 0 to 50, as found in the GPLURAL, DIMIN, PP, and CHUNK data sets.

training examples. In Figure 6.6 we visualize the outcome of one analysis we performed on the GPLURAL, DIMIN, PP, and CHUNK tasks. The scatter plots in Figure 6.6 relate each example's CPS as measured in a leave-one-out experiment on training material (x-axis) to its actual CPS as measured on the test data (y-axis). Each point represents one or more examples. As all four scatter plots show, there is an apparent rough correlation between the two, but also a considerable amount of deviation from the $x = y$ diagonal. This means that, interestingly, test data only partly follow the class borders present in the training set. Or alternatively put, the test data contain a lot of counter-evidence to the class areas assumed by the training examples, for all four tasks displayed. Sometimes low-CPS examples (bad neighbors) in the training data become excellent predictors of their class in test data; alternatively, high-CPS examples can be bad predictors of new test instances' classes.

In other words, bad neighbors or exceptional examples in the training data do not necessarily lead to classification errors in test data. A second insightful quantitative analysis of the relation between CPS and actual classification behavior on test data is visualized for the GPLURAL, DIMIN,

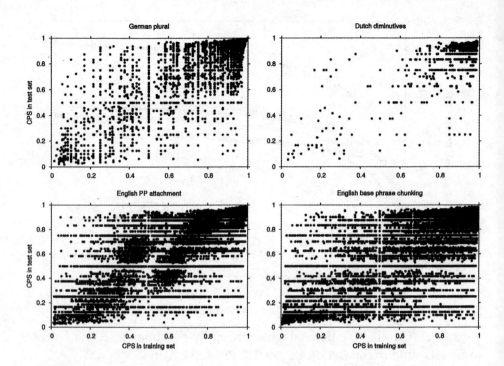

Figure 6.6: Scatter plots of CPS values of training examples when measured on a leave-one-out experiment on training data (horizontal) and when classifying test data (vertical) on four NLP tasks.

PP, and CHUNK tasks in Figure 6.7. For each of the four tasks two surfaces are plotted in a three-dimensional space, where one surface represents all 1-nearest neighbors of test instances that predict the correct class (labeled "friendly"), and the other surface represents all remaining 1-nearest neighbors that predict the incorrect class (labeled "unfriendly"). For all nearest neighbors, their CPS in the training set is related to their actual distance to the test instance on the x and y axes. The actual numbers of friendly and unfriendly nearest neighbors were aggregated in a 10×10 matrix, each dimension represented by ten equal-width bins of normalized distance and CPS, respectively. In Figure 6.7 the z axis represents these bare counts in a logarithmic scale.

If low-CPS examples would always cause misclassifications, then the "unfriendly" surfaces would be positioned above the "friendly" surfaces in the front areas of the four surface plots, at the lower values of CPS. In the experiments on the four tasks, however, the "unfriendly" surfaces

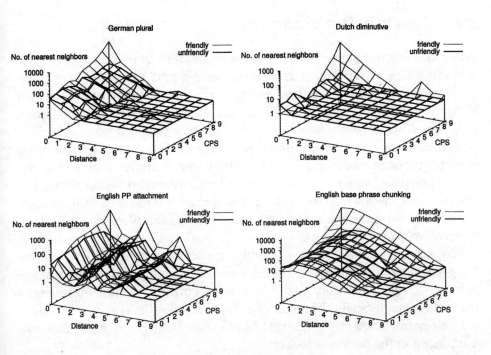

Figure 6.7: Surface plots of the number of nearest neighbors (z-axis; log-scale) as a function of CPS (counted in 10% CPS intervals) and normalized distance of the nearest neighbor. Each plot contains a separate surface for friendly neighbors (predicting a correct class) and unfriendly neighbors (predicting an incorrect class).

are positioned *under* the friendly surfaces except slightly in the far corner representing the cases with the lowest CPS. In these areas there are also still substantial amounts of low-CPS examples that predict the correct class.

In other words, misclassifications hardly ever outnumber correct classifications at any interval, except at the lowest values of CPS. In fact, most examples at the lower half of the CPS spectrum (below 0.5) cause correct classifications. This is a significant finding – although relatively more classification errors are due to low-CPS or "exceptional" examples, it is not the case that the use of low-CPS examples as nearest neighbors exclusively leads to errors - on the contrary. This explains why removing low-CPS examples does not improve (or hardly improves) generalization accuracy, and tends to lead to lower accuracies when more examples are edited.

6.4 Generalizing examples

While keeping full memory may be a safe guideline to avoid any eventual harmful effect of editing, it is still interesting and tempting to explore other means to reduce the need for memory, provided that performance is not harmed. As we have illustrated in the previous section, quite large amounts of examples can be forgotten while retaining a similar performance as the memory-based learner with full memory. The point at which performance starts degrading, however, is unpredictable from the results presented so far. In this section we explore methods that attempt to abstract over memorized examples in a different and more careful manner, namely by merging examples into generalized examples, using various types of merging operations.

We start, in subsection 6.4.1, with an overview of existing methods for generalizing examples in memory-based learning. Subsequently we present FAMBL, a memory-based learning algorithm variant that merges same-class nearest-neighbor examples into "families". In subsection 6.4.3 we compare FAMBL to pure IB1 on the same range of NLP tasks as introduced in the previous section.

6.4.1 Careful abstraction in memory-based learning

Early work on the k-NN classifier pointed at advantageous properties of the classifier in terms of generalization accuracies, under certain assumptions, because of its reliance on full memory (Fix & Hodges, 1951; Cover & Hart, 1967). However, the trade-off downside of full memory is computational inefficiency of the classification process, as compared to classifiers that do abstract from the learning material. Therefore, a part of the early work in k-NN classification focused on *editing* methods, as touched upon in the previous section.

The renewed interest in the k-NN classifier from the late 1980s onwards in the AI subfield of machine learning (Stanfill & Waltz, 1986; Stanfill, 1987; Aha et al., 1991; Salzberg, 1991) caused several new implementations of ideas on criteria for editing, but also other approaches to abstraction in memory-based learning emerged. In this subsection we present FAMBL, a carefully-abstracting memory-based learning algorithm. FAMBL merges groups of very similar examples (families) into family expressions.

6.4. GENERALIZING EXAMPLES

Carefully merged examples

Paths in decision trees can be seen as generalized examples. In IGTREE and C4.5 (Quinlan, 1993) this generalization is performed up to the point where no actual example is left in memory; all is converted to nodes and arcs. Counter to this decision-tree compression, approaches exist that start with storing individual examples in memory, and carefully merge some of these examples to become a single, more general example, only when there is some evidence that this operation is not harmful to generalization performance. Although overall memory is compressed, the memory still contains individual items on which the same k-NN-based classification can be performed. The abstraction occurring in this approach is that after a merge, the merged examples incorporated in the new generalized example are deleted individually, and cannot be reconstructed. Example approaches to merging examples are NGE (Salzberg, 1991) and its batch variant BNGE (Wettschereck & Dietterich, 1995), and RISE (Domingos, 1996). We provide brief discussions of two of these algorithms: NGE and RISE.

NGE (Salzberg, 1991), an acronym for *Nested Generalized Exemplars*, is an incremental learning theory for merging examples (or exemplars, as Salzberg prefers to refer to examples stored in memory) into *hyperrectangles*, a geometrically motivated term for merged exemplars. NGE[2] adds examples to memory in an incremental fashion (at the onset of learning, the memory is seeded with a small number of randomly picked examples). Every time a new example is presented, it is matched with all exemplars in memory, which can be individual or merged exemplars (hyperrectangles). When it is classified correctly by its nearest neighbor (an individual exemplar or the smallest matching hyperrectangle), the new example is merged with it, yielding a new, more general hyperrectangle.

Figure 6.8 illustrates two mergings of examples of the German plural task with exemplars. On the top of figure 6.8, the example -urSrIftF (from the female-gender word Urschrift), labeled with class en (representing the plural form Urschriften), is merged with the example t@rSrIftF (from the female-gender word Unterschrift), also of class en, to form the generalized exemplar displayed on the right-hand side. On the first two features, a disjunction is formed of, respectively, the values - and t, and u and @. This means that the generalized example matches on any other example that has value - *or* value t on the first feature, and any other example that has value u *or* value @ on the second feature.

---

[2]Salzberg (1991) makes an explicit distinction between NGE as a theory, and the learning algorithm EACH as the implementation; we will use NGE here to denote both.

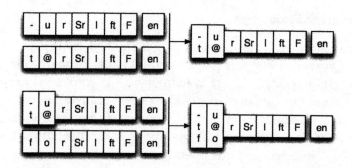

Figure 6.8: Two examples of the generation of a new hyperrectangle in NGE: from a new example and an individual exemplar (top) and from a new example and the hyperrectangle from the top example (bottom).

The lower part of Figure 6.8 displays a subsequent merge of the newly generalized example with another same-class example, forSrIftF (the female-gender word Forschrift), which leads to a further generalization of the first two features.

In nested generalized examples, abstraction occurs because it is not possible to retrieve the individual examples nested in the generalized example; new generalization occurs because the generalized example not only matches fully with its nested examples, but would also match perfectly with potential examples with feature-value combinations that were not present in the nested examples; the generalized example in Figure 6.8 would also match torSrIft, f@rSrIft, furSrIft, -orSrIft. These examples do not necessarily match existing German words, but they might – and arguably they would be labeled with the correct plural inflection class.

RISE (*Rule Induction from a Set of Exemplars*) (Domingos, 1995; Domingos, 1996) is a multi-strategy learning method that combines memory-based learning with rule-induction (Michalski, 1983; Clark & Niblett, 1989; Clark & Boswell, 1991). As in NGE, the basic method is that of a memory-based learner and classifier, only operating on a more general type of example. RISE learns a memory filled with *rules* which are all derived from individual examples. Some rules are example-specific, and other rules are generalized over sets of examples.

RISE inherits parts of the rule induction method of CN2 (Clark & Niblett, 1989; Clark & Boswell, 1991). CN2 is an incremental rule-induction algorithm that attempts to find the "best" rule governing a certain amount of examples in the example base that are not yet covered by a rule. "Goodness" of a rule is estimated by computing its apparent accuracy

6.4. GENERALIZING EXAMPLES

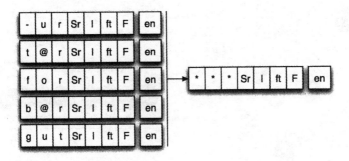

Figure 6.9: An example of an induced rule in RISE, displayed on the right, with the set of examples that it covers (and from which it was generated) on the left.

(which is class prediction strength, Cost & Salzberg, 1993) with Laplace correction (Niblett, 1987; Clark & Boswell, 1991).

RISE induces rules in a careful manner, operating in cycles. At the onset of learning, all examples are converted to example-specific rules. During a cycle, for each rule a search is made for the nearest example not already covered by it that has the same class. If such an example is found, rule and example are merged into a more general rule. Instead of disjunctions of values, RISE generalizes by inserting wild card symbols (that match with any other value) on positions with differing values. At each cycle, the goodness of the rule set on the original training material (the individual examples) is monitored. RISE halts when this accuracy measure does not improve (which may already be the case in the first cycle, yielding a plain memory-based learning algorithm).

Figure 6.9 illustrates the merging of individual examples into a rule. The rule contains seven normally valued conditions, and two wild cards, '*'. The rule now matches on every female-gender example ending in Srift (Schrift).

RISE classifies new examples by searching for the best-matching rule, always selecting the rule with the highest Laplace accuracy (Clark & Boswell, 1991). As a heuristic add-on for dealing with symbolic values, RISE incorporates a value-difference metric (Stanfill & Waltz, 1986; Cost & Salzberg, 1993) by default, called the *simplified value-difference metric* (SVDM) due to its simplified treatment of feature-value occurrences in the VDM function (Domingos, 1996).

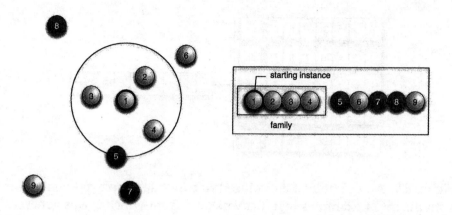

Figure 6.10: An example of a family in a two-dimensional example space (left). The family, at the inside of the circle, spans the focus example (marked with number 1) and the three nearest neighbors labeled with the same class (indicated by their color). When ranked in the order of distance (right), the family boundary is put immediately before the first example of a different class, the gray example with number 5.

FAMBL: merging example families

FAMBL, for *FAMily-Based Learning*, is a variant of IB1 that constitutes an alternative approach to careful abstraction over examples. The core idea of FAMBL, in the spirit of NGE and RISE, is to transform an example base into a set of *example family expressions*. An example family expression is a hyperrectangle, but the procedure for merging examples differs from that in NGE or in RISE. First, we outline the ideas and assumptions underlying FAMBL. We then give a procedural description of the learning algorithm.

Classification of an example in memory-based learning involves a search for the nearest neighbors of that example. The value of k in k-NN determines how many of these neighbors are used for extrapolating their (majority) classification to the new example. A fixed k ignores (smoothes) the fact that an example is often surrounded in example space by a number of examples of the same class that is actually larger or smaller than k. We refer to such a variable-sized set of same-class nearest neighbors as an example's *family*. The extreme cases are on the one hand examples that have a nearest neighbor of a different class, i.e., they have no family members and are a family on their own (and have a CPS of 0.0, cf. section 6.1), and on the other hand examples that have as nearest neighbors all other examples of the same class.

6.4. GENERALIZING EXAMPLES

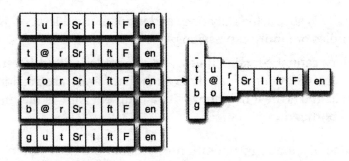

Figure 6.11: An example of family creation in FAMBL. Five German plural examples (left) are merged into a family expression (right).

Thus, families represent same-class clusters in example space, and the number and sizes of families in a data set reflect the *disjunctivity* of the data set: the degree of scatteredness of classes into clusters. In real-world data sets, the situation is generally somewhere between the extremes of total disjunctivity (one example per cluster) and no disjunctivity (one cluster per class). Many types of language data appear to be quite disjunct (Daelemans et al., 1999), and we have provided an analysis of the disjunctivity of the six benchmark tasks used in this chapter in section 6.3 (in particular, cf. Figure 6.5) that illustrates this fact. In highly disjunct data, classes are scattered among many small clusters, which means that examples have few nearest neighbors of the same class on average.

Figure 6.10 illustrates how FAMBL determines the family of an example in a simple two-dimensional example space. All nearest neighbors of a randomly picked starting example (marked by the black dot) are searched and ranked in the order of their distance to the starting example. Although there are five examples of the same class in the example space, the family of the starting example contains only three examples, since its fourth-nearest example is of a different class.

Families are converted in FAMBL to *family expressions*, which are hyper-rectangles, by merging all examples belonging to that family simultaneously. Figure 6.11 illustrates the creation of a family expression from an example family. In contrast with NGE,

- family expressions are created in one non-incremental operation on the entire example base, rather than by step-wise nesting of each individual family member;

- a family is abstracted only once and is not merged later on with other examples or family expressions;
- families cannot contain "holes", i.e., examples with different classes, since the definition of family is such that family abstraction halts as soon as the nearest neighbor with a different class is met in the local neighborhood.

The general modus of operation of FAMBL is that it randomly picks examples from an example base one by one from the set of examples that are not already part of a family. For each newly picked example, FAMBL determines its family, generates a family expression from this set of examples, and then marks all involved examples as belonging to a family (so that they will not be picked as a starting point or member of another family). FAMBL continues determining families until all examples are marked as belonging to a family.

Families reflect the locally optimal k surrounding the example around which the family is created. The locally optimal k is a notion that is also used in locally weighted learning methods (Vapnik & Bottou, 1993; Wettschereck & Dietterich, 1994; Wettschereck, 1994; Atkeson et al., 1997); however, these methods do not abstract from the learning material. In this sense, FAMBL can be seen as a local abstractor.

The FAMBL algorithm converts any training set of labeled examples to a set of family expressions, following the procedure given in Figure 6.12. In essence, FAMBL continuously selects a random example, and extends it to a family expression, until all examples are captured in a family. After learning, the original example base is discarded, and further classification is based only on the set of family expressions yielded by the family-extraction phase. Classification in FAMBL works analogously to classification in pure memory-based learning (with the same similarity and weighting metrics as we used so far with MBL): a match is made between a new test example and all stored family expressions. When a family expression contains a disjunction of values for a certain feature, a match is counted when one of the disjunctive values matches the value at that feature in the new example. How the match is counted exactly depends on the similarity metric. With the overlap metric, the feature weight of the matching feature is counted, while with the MVDM metric the smallest MVDM distance among the disjuncted feature values is also incorporated in the count.

6.4. GENERALIZING EXAMPLES

Procedure FAMBL FAMILY-EXTRACTION:

Input: A training set TS of examples $I_{1...n}$, each example being labeled with a family-membership flag set to *FALSE*

Output: A family set FS of family expressions $F_{1...m}$, $m \leq n$

$i = f = 0$

1. Randomize the ordering of examples in TS
2. While not all family-membership flags are *TRUE*, Do
    - While the family-membership flag of I_i is *TRUE* Do increase i
    - Compute NS, a ranked set of nearest neighbors to I_i with the same class as I_i, among all examples with family-membership flag *FALSE*. Nearest-neighbor examples of a different class with family-membership flag *TRUE* are still used for marking the boundaries of the family.
    - Set the membership flags of I_i and all remaining examples in NS to *TRUE*
    - Merge I_i and all examples in NS into the family expression F_f and store this expression along with a count of the number of example merged in it
    - $f = f + 1$

Figure 6.12: Pseudo-code of the family extraction procedure in FAMBL.

6.4.2 Getting started with FAMBL

FAMBL has been implemented as a separate software package. It offers the Fambl executable which offers the same similarity and weighing functions as TIMBL. Its output in the terminal where FAMBL is run, however, provides some additional statistical information on the families extracted.

On the German plural data set introduced in chapter 3, the basic FAMBL command to be issued would be the following:

```
% Fambl -f gplural.train -t gplural.test
```

```
Fambl (Family-based learning), version 2.2.1, 1 November 2004
(c) 1997-2004 ILK Research Group, Tilburg University
http://ilk.uvt.nl / antalb@uvt.nl
current time: Thu Nov 11 20:36:05 2004
    metric scheme set to GR feature wgts, no distance wgts, MVDM
    sorting and reading data base gplural.train
    data base has 12584 instances
                  7701 instance types
                     7 features
                     8 classes
    computing feature gain ratio values
    feature  0 (   73 values):  0.026694
    feature  1 (   27 values):  0.048522
    feature  2 (   77 values):  0.062067
    feature  3 (   81 values):  0.065165
    feature  4 (   24 values):  0.218643
    feature  5 (   74 values):  0.204404
    feature  6 (    8 values):  0.463801
    average gain ratio: 0.155614
    presorting and rereading instance base gplural.train
    computing MVDM matrices
    took 0 seconds
<*> family extraction stage started
    # families                             :      3435
    average # members                      :    3.6635
    average k distances                    :    1.8908
    average description length (bytes)     :   77.8748
    compression (raw memory)               :   33.5714 %
    compression (vs instance types)        :    3.5118 %
    #type vs. #fam reduction               :   55.3954 %
    clusteredness                          :    817.72
    took 6 seconds
```

After reading the training examples, FAMBL commences its family extraction phase, reporting on the progress, and upon completion, reporting on the statistics of the extracted family expressions. This particular FAMBL run compresses the German plural training set of 12,584 examples into 3,163 family expressions, with an average membership of about 4 examples, spread over about 2 distances on average. The average family in the current implementation takes about 80 bytes to store, which leads to a memory compression rate of about 37%, and 59% compression in the number of family expressions as compared with the original number of example types.

Subsequently, FAMBL classifies the test set, reporting on progress and, upon completion, reports on the percentage of correctly classified test instances. An output file is created of which the name follows a syntax similar to that of the output files of TIMBL.

6.4. GENERALIZING EXAMPLES

The current example introduces one new symbol, K, which is the number of distances that can maximally be taken into account in family creation. The default value of K is 3.

```
<*> starting test with k=1
    writing output to gplural.test.Fambl.M.gr.k1.K3.out
      1000 instances processed,   93.5000 % correct
      2000 instances processed,   93.6000 % correct
     ....
     12000 instances processed,   93.8500 % correct
     11803 instances out of    12584 classified correctly
    Fambl score:  93.7937 % correct instances
    took 18 seconds
    (699.11 instances per second)
<*> current time: Thu Nov 11 20:36:29 2004
    Fambl spent a total of 24 seconds running;
    6 on learning, 18 on testing.
    Fambl ready.
```

After it has classified the test data, FAMBL closes with reporting on the percentage of correctly classified test instances, and the number of seconds (elapsed wall clock time) that were involved in classification.

6.4.3 Experiments with FAMBL

We performed experiments with FAMBL on the same six language processing tasks used earlier in this chapter with the editing experiments. As a first experiment, we varied both the normal k parameter (which sets the number of distances in the nearest neighbor set used in k-NN classification), and the FAMBL-specific parameter that sets the maximum k distances in the family extraction stage, which we will refer to as K (the -K parameter in the command-line version of FAMBL). The two parameters are obviously related - the K can be seen as a preprocessing step that "precompiles" the k for the k-NN classifier. The k-NN classifier that operates on the set of family expressions can be set to 1, hypothetically, since the complete example space is pre-partitioned in many small regions of various sizes (with maximally K different distances) that each represent a locally appropriate k.

If the empirical results would indeed show that k can be set to 1 safely when K is set at an appropriately large value, then FAMBL could be seen as a means to factor the important k parameter away from IB1. We performed comparative experiments with IB1 and FAMBL on the six benchmark tasks, in which we varied both the k parameter in IB1, and the K parameter in FAMBL while keeping $k = 1$. Both k and K were varied in the pseudo-

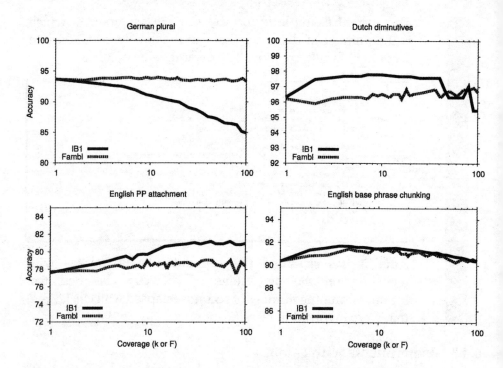

Figure 6.13: Generalization accuracies (in terms of % of correctly classified test instances) and F-scores, where appropriate, of IB1 with increasing k parameter, and FAMBL with $k = 1$ and increasing K parameter.

exponential series $[0, 1, \ldots, 9, 10, 15, \ldots, 45, 50, 60, \ldots, 90, 100]$. The results of the experiments are visualized in Figure 6.13.

A very large value of K means that FAMBL incorporates virtually any same-class nearest neighbor at any furthest distance in creating a family, as long as there are no different-class nearest neighbors in between. It would be preferable to be able to fix K at a very high value without generalization performance loss, since this would effectively factor out not only the k parameter, but also the K parameter. This situation is represented quite perfectly in the graph displaying the results of GPLURAL (top left corner of Figure 6.13). While a larger k in IB1 leads to a steady decline in generalization accuracy on test data of the GPLURAL task, FAMBL's accuracy remains very much at the same level regardless of the value of K. The results with the other three tasks also show a remarkably steady generalization accuracy (or F-score, with CHUNK) of FAMBL, with increasing K, but in all three cases FAMBL's score is not higher than IB1's.

6.4. GENERALIZING EXAMPLES

Especially with the DIMIN and PP tasks, matching on families rather than on examples leads to less accurate classifications at wide ranges of K.

The k parameter can, of course, be set at $k > 1$ in FAMBL as well. Classification with $k > 1$ in FAMBL means that the k-nearest families are sought to base classification on. A larger k might help in smoothing problematic cases in which a single small nearest neighbor family is surrounded by different-class families, and a new test instance appears at the border of the two class areas. For three tasks, GPLURAL, DIMIN, and PP, we exhaustively ran experiments with increasing K and increasing k, where we varied k in the range $[1\ldots10]$. Figure 6.14 displays the generalization accuracies of FAMBL as the third axis in a three-dimensional surface plot. The bottom of the graph displays in grayscale the relative height of each of the points on the surface grid above it.

Figure 6.14 essentially shows three fairly flat surfaces. If they would have been entirely flat, then we would have seen strong empirical support for the claim that both the K and k parameter are effectively factored out by FAMBL, because the same (high) performance can be obtained with any combination of values. Also, recall that the generalization accuracies were close to (or better than, with GPLURAL) IB1's. What we see, however, is that with GPLURAL, the surface has two peaking ridges of best scores along the x and y axes, i.e., at $K = 1$ and at $k = 1$; any combination of $K > 1$ and $k > 1$ lead to (slightly) lower generalization accuracies. With DIMIN, the best accuracies occur with $k > 1$ and $K < 10$. With PP, slightly better accuracies are obtained with higher values of k and K. These surfaces appear to follow the trends of the IB1 curves in Figure 6.13, which show minor improvements with $k = 10$ over $k = 1$.

While it retains a similar performance to IB1, FAMBL also attains a certain level of compression. This can be measured in at least two ways. First, in Figure 6.15 the amount of compression (in terms of percentages) is displayed of the number of families versus the original number of examples, with increasing values of K, for four of our tasks. As Figure 6.15 shows, the compression rates converge for all four tasks at similar and very high levels; between 77% for GPLURAL to 92% for DIMIN. Apparently, setting K at a large enough value ensures that at that point even the largest families are identified; typically there will be 100 or less different distances in any found family.

Some more detailed statistics on family extraction are listed in Table 6.5, measured for four tasks at the $K = 100$ mark. The actual number of families varies widely among the tasks, but this correlates with the number of training examples (cf. Table 6.1). The average number of members

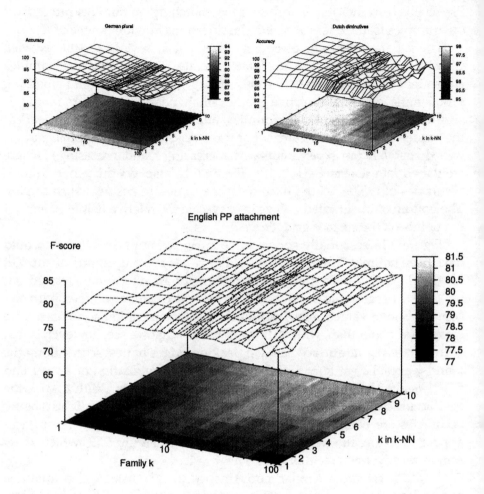

Figure 6.14: Generalization accuracies (F-score %) by FAMBL on the GPLURAL, DIMIN, and PP tasks, plotted as a function of the K parameter (maximal number of distances in one family) and the k parameter (of the k-NN classifier).

lies at about the same order of magnitude for the four tasks – between six and thirteen. The table also shows the raw memory compression when compared with a straightforward storage of the flat example base. In the straightforward implementation of FAMBL, storing a family with

6.4. GENERALIZING EXAMPLES

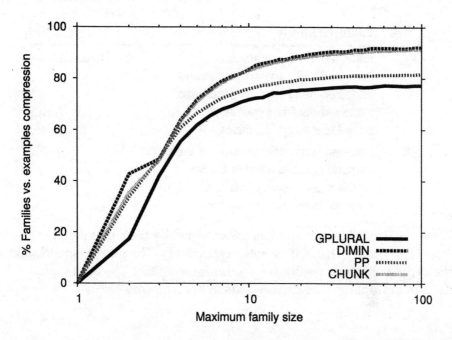

Figure 6.15: Compression rates (percentages) of families as opposed to the original number of examples, produced by FAMBL at different maximal family sizes (represented by the x-axis, displayed at a log scale).

Task	Number of families	Av. number of members	Memory compression (%)
GPLURAL	1,749	7.2	62.0
DIMIN	233	12.9	73.4
PP	3,613	5.8	23.4
CHUNK	17,984	11.8	51.9

Table 6.5: Number of extracted families at a maximum family size of 100, the average number of family members, and the raw memory compression, for four tasks.

one example uses more memory than storing one example because of the bookkeeping information associated with storing possible disjunctions at each feature. The net gains of the high compression rates displayed in Figure 6.15 are still positive: from 23% to 73% compression. This is, however, dependent on the particular implementation.

Task	Example family	Class
PP	attributed gains to demand	Verb attachment
	attributed improvement to demand	
	attributed performance to increases	
	attributed decline to demand	
	bring focus to opportunities	
NP	because computers do **most** of the work	B-NP
	demand rights to **most** of the 50	
	he still makes **most** of his furs	
	screens , said **most** of the top	

Table 6.6: Two example families (represented by their members) extracted from the PP and CHUNK data sets, respectively. The part-of-speech tags in the CHUNK example family are left out for legibility. The bold words in the CHUNK example are the focus words in the windows.

Two example families, one for the PP and the other for the CHUNK task, are displayed in Table 6.6. The first example family, labeled with the Verb attachment class, represents the attributed ... to ... pattern, but also includes the example bring focus to opportunities, which is apparently the closest neighbor to the other four examples having the same class. The second family represents cases of the beginning of a noun phrase starting with most of. The context left of most of deviates totally between the four examples making up the family, while the right context represents a noun phrase beginning with the or his. This family would also perfectly match sentence fragments inside the family hyperrectangle, such as because computers do most of the top, or he still makes most of the 50, and many more recombinations. Analogously, the PP family example displayed in Table 6.6 would also perfectly match attributed decline to increases, bring focus to demand, etcetera.

Overall, the comparison between FAMBL and IB1 shows that FAMBL does not profit from the relatively large generalizing capacity of family expressions, that in principle would allow some unseen examples attain a higher score in the similarity function. Apart from the question whether this relative re-ranking of examples would have any effect on classification, it is obvious that many examples covered by family expressions are unlikely to occur — consider, for example, because computers do most of his furs.

6.5. CONCLUSION

We conclude that FAMBL has two main merits. First, FAMBL can compress an example base down to a smaller set of family expressions (or, a generalizing hyperrectangle), attaining various compression rates in the same ballpark as attained by editing methods, but with a very steady generalization accuracy that is very close to IB1's. Second, FAMBL almost factors the k parameter out. Fairly constant performance was observed while keeping $k = 1$ and varying K, the maximal number of family members, across a wide range of values. To sum up, FAMBL is a successful local k pre-compiler.

The relation between FAMBL and editing based on CPS is that FAMBL actively preserves class borders, since these borders by definition mark the outer edges of the hyperrectangles that FAMBL builds. Arguably, by being more careful about class borders, FAMBL is more successful as an abstraction method than CPS-based editing. Success, however, is constrained in both cases to compression; abstraction does not lead to performance gains.

6.5 Conclusion

Memory-based learning is the opposite of what the minimal description length (MDL) principle recommends a learning algorithm to be. It does not search for a minimally-sized model of the classification task for which examples are available. Rather, it stores all examples in memory — it settles for a state of maximal description length. This extreme bias makes memory-based learning an interesting comparative case against so-called eager learners, such as decision tree induction algorithms and rule learners, that do adhere to the MDL. If the MDL is right, then the hypothesis is that memory-based learning produces less optimal models, in terms of generalization performance, than eager learners do.

This hypothesis is too strong; eager learners are not entirely faithful to the MDL. The definition of MDL discerns between the size of the abstracted model on the one hand, and the size of the exception list on the other hand that is needed to capture a recurring set of examples that do not fit the abstracted model (Rissanen, 1983). Eager learners typically only focus on inducing an abstracted model and do not produce an exception list to go with the model. In inducing the abstracted model, eager learners filter out noise, usually by searching for low-frequency events and events that seem to contradict their local neighborhood in the feature space according to tests (e.g., statistical tests using the binomial distribution). The classical C4.5

algorithm offers two pruning parameters that filter noise in these two ways (Quinlan, 1993). Indeed, if noise is defined as a random source that can freely change feature values and class labels into incorrect values, then by that definition it is not only safe, but also a good idea to ignore noise. It is wise not to waste any space in the abstracted model on them, and it is also unnecessary to store them in an exception list, since random noise will not reoccur.

Indeed, data sets of many real-world tasks contain noise, and decision-tree pruning techniques such as C4.5 (Quinlan, 1993) and rule pruning techniques such as RIPPER (Cohen, 1995) are known to be succesful noise filters. But also in memory-based learning noise filtering is an issue. In their seminal paper on instance-based learning, Aha et al. (1991) describe the two instance-based learners IB2 and IB3, that both actively filter out noise by editing examples. Due to its reliance on local neighborhood, the k-NN classifier is extremely sensitive to noise, especially with $k = 1$ (Wilson, 1972; Tomek, 1976).

Yet, our results contradict the assumption that eager learners are better than lazy learners. The results show that IB1 outperforms RIPPER on six NLP tasks, often by wide margins. Earlier we compared IB1 to C4.5 (Daelemans et al., 1999), attaining an equal performance on one task (English part-of-speech tagging), but yielding significantly better performance of IB1 over C4.5 on English grapheme-phoneme conversion and PP-attachment.

This contradiction has a two-fold explanation. First, the NLP data on which we base our studies hardly contain noise. The data sets we use, the Penn Treebank and CELEX-2, were very carefully annotated and checked for errors. Second, all six language data sets contain many individual examples that have deviating class labels from their nearest neighbors, as for example Figure 6.5 showed, so they can be said to contain a large amount of exceptions that can recur eventually in new data. On the basis of this we conclude that any algorithm that decides to discard low-frequency examples dissimilar to their local neighborhood is at a disadvantage to language data as compared to memory-based learning.

Our editing experiments furthermore brought to light that editing never clearly improved generalization results; it largely harmed them when larger amounts of examples were edited, no matter by what criterion. Most damage appeared to be inflicted by removing examples that could be likened to support vectors from support vector machines, i.e., examples close to a hyperplane border between classes; the least damage was inflicted by randomly removing examples, making the memory gradually but uniformly more sparse. Closer analyses showed that many examples with a low CPS actually turned out very good predictors when they were used as nearest neighbors, while many other examples with a high CPS turned out bad predictors for new data. If anything, this indicates that it does not make sense to uphold a division of examples into exceptional and regular ones, since both may recur in new data and both may be good neighbors. Indeed, memory-based learning does not make this division and treats all examples equally, as if they form one big exception list and no model. Arguably, the MDL principle may not be violated at all here — for NLP data it might mean that there is simply not much that can be abstracted or minimized without weakening the model.

Our experiments with FAMBL show that some compression of the examples in memory can in fact be attained, not by editing examples, but rather by forming generalized examples (families) that uphold the original class borders in the data. Some example-specific information is lost in this process, but judging from the results it appears to be important that the class borders within the data are maintained — FAMBL displays very stable generalization performance scores close to IB1's. An additional advantage of FAMBL is that it appears to render the k parameter redundant; with $k = 1$, families can be formed with a wide range of limits on numbers of family members without considerable deviations from generalization performance. Still, all the effort put in abstraction from examples to families in FAMBL does not lead to performance gains, just as editing failed to achieve. We have quite some evidence now that abstracting from data is not helpful in learning NLP tasks in terms of optimal generalization performance; only when interested in speed gains and memory-lean models, at the cost of (sometimes mild) performance losses, do abstracting methods such as FAMBL, IGTREE, and TRIBL , and eager learners such as RIPPER and C4.5 offer opportunities.

6.6 Further reading

This chapter is a reconstruction and extension with more analysis and slightly different results of research presented earlier in Daelemans et al. (1999) on the harmfulness of forgetting exceptions. Initial work with FAMBL is described in Van den Bosch (1999). The issue of the harmfulness of forgetting has also been in focus in the context of probabilistic models of natural language processing. The ruling hypothesis in the development of probabilistic models has for a long time been that what is exceptional (improbable) is unimportant. For about a decade, however, empirical evidence has been gathering in research on probabilistic models that supports the "forgetting is harmful" hypothesis. For instance, in Bod (1995), a data-oriented approach to parsing is described in which a treebank is used as a memory and in which the parse of a new sentence is computed by reconstruction from subtrees present in the treebank. It is shown that removing all hapaxes (unique subtrees) from memory degrades generalization performance from 96% to 92%. Bod notes that "this seems to contradict the fact that probabilities based on sparse data are not reliable" (Bod, 1995, p.68). In the same vein, Collins and Brooks (1995) show that when applying the back-off estimation technique (Katz, 1987) to learning prepositional-phrase attachment, removing all events with a frequency of less than 5 degrades generalization performance from 84.1% to 81.6%. In Dagan et al. (1997), a similarity-based estimation method is compared to back-off and maximum-likelihood estimation on a pseudo-word sense disambiguation task. Again, a positive effect of events with frequency 1 in the training set on generalization accuracy is noted.

A direct comparison between lazy and eager learning applied to natural language processing tasks, as well as a commentary on (Daelemans et al., 1999) can be found in (Rotaru & Litman, 2003), who perform editing experiments with IB1, and compare IB1 versus RIPPER on four understanding tasks in spoken dialog systems. Next to class prediction strength, Rotaru and Litman also use *typicality* (Zhang, 1992), a global measure of exceptionality, and *local typicality*, a local version of the typicality metric, which expresses the average similarity of a training example to nearest neighbors with the same class divided by the average similarity of the example to nearest neighbors with a different class. Editing experiments on one task show moderate improvements with some editing methods at 10-50% editing rates, especially when editing examples with a high local typicality. Rotaru and Litman also use the three editing criteria to compare the generalization performances of IB1 and RIPPER, and observe a trend that IB1 is slightly better than RIPPER in classifying examples with a below-0.5 CPS or local typicality.

The early work on editing (Hart, 1968; Wilson, 1972) has triggered many additional investigations in pattern recognition research (Swonger, 1972; Tomek, 1976; Devijver & Kittler, 1980; Wilson & Martinez, 1997). We particularly mention Brighton and Mellish (2002) who describe the Iterative Case Filtering (ICF) filtering algorithm that actively preserves border or "support vector" instances. Brighton and Mellish (2002) note that the CPS criterion used here and in (Daelemans et al., 1999) does not ensure the proper retention of these border cases. Our results suggest this is indeed the case; FAMBL can be seen as a more successful method to attain compression while retaining all class borders.

Chapter 7

Extensions

This chapter describes two complementary extensions to memory-based learning: a search method for optimizing parameter settings, and methods for reducing the near-sightedness of the standard memory-based learner to its own contextual decisions in sequence processing tasks. Both complement the core algorithm as we have been discussing so far. Both methods have a wider applicability than just memory-based learning, and can be combined with any classification-based supervised learning algorithm.

First, in section 7.1 we introduce a search method for finding optimal algorithmic parameter settings. No universal rules of thumb exist for setting parameters such as the k in the k-NN classification rule, or the feature weighting metric, or the distance weighting metric. They also interact in unpredictable ways. Yet, parameter settings do matter; they can seriously change generalization performance on unseen data. We show that applying heuristic search methods in an experimental wrapping environment (in which a training set is further divided into training and validation sets) can produce good parameter settings automatically.

Second, in section 7.2 we describe two technical solutions to the problem of "sequence near-sightedness" from which many machine-learning classifiers and stochastic models suffer that predict class symbols without coordinating one prediction with another in some way. When such a classifier is performing natural language sequence tasks, producing class symbol by class symbol, it is unable to stop itself from generating output sequences that are impossible and invalid, because information on the output sequence being generated is not available to the learner. Memory-based learning is no exception, but there are remedies which we describe in section 7.2. As one remedy extension we already presented MBT, which creates a feedback loop between the classifier's output and its input, so that it can base its next classification partly on its previous ones. Another remedy, presented in section 7.2.1, is to stack classifiers – i.e., to have a second classifier learn from examples that are enriched by the classifications of a first near-sighted classifier. A third solution, described in section 7.2.2 is to predict sequences of class symbols rather than to predict single class symbols. We test the two methods on sequence tasks that have been investigated in the previous chapters: Dutch morphological analysis (MORPH), English base phrase chunking (CHUNK), and named-entity recognition (NER). We test stacking and sequence prediction on these three tasks, and also show the combination of both on these tasks. The results are striking; both classifier stacking and predicting sequences of class symbols lead to higher F-scores on all three tasks, and the best improvements are obtained by combining the two methods.

7.1 Wrapped progressive sampling

It is common knowledge that large changes can be observed in the generalization accuracy of a machine learning algorithm on some task when instead of its default algorithmic parameter settings, one or more parameters are given a non-default value. This is definitely the case with IB1, which for example, as we exemplified on several occasions earlier in this book, is highly sensitive to changing the value of the k parameter. The k parameter in turn influences the effects that distance weighting and the use of the MVDM kernel can have. Another range of examples is given in (Daelemans & Hoste, 2002). Differences in accuracy on some task due to different parameter settings of the same algorithm can easily overwhelm accuracy differences between two different algorithms using default settings for algorithm parameters. The fundamental problems in algorithmic parameter selection (or model selection) are that it is hard to estimate which parameter setting would lead to optimal generalization performance, and that this estimation has to be redone for each task. One can estimate it on the labeled data available for training purposes, but optimizing parameters on training material easily leads to overfitting. A remedy for overfitting is to use classifier wrapping (Kohavi & John, 1997), which partitions the available labeled training material into internal training and test data, and which performs cross-validation experiments to estimate a training-set-internal generalization accuracy. Using this method, competitions can be held among parameter settings, to determine the average best-performing setting to be used later in the experiment on the real test data.

For many tasks it is not feasible to test all possible combinations of parameter settings exhaustively. To allow the vast search space of possible parameter setting combinations to be sufficiently accessible, search methods can come to our aid. Here we describe such a method. It is based on wrapped progressive sampling, which borrows its basic heuristic from progressive sampling (Provost et al., 1999). The goal of progressive sampling is to perform iterative experiments using a growing data set, and halt the growing at the point at which generalization performance on held-out validation material does not improve as compared to the previous steps. In our parameter optimization process we do not adopt this convergence goal, but we do adopt the progressive sampling method in which we test *decreasing* amounts of combinations of settings with *increasing* amounts of training data – inheriting the favorable speedups of progressive sampling (Provost et al., 1999).

7.1.1 The wrapped progressive sampling algorithm

The wrapped progressive sampling (henceforth WPS) algorithm, takes as input a data set of labeled examples D, and produces as output a combination of parameter settings that is estimated to lead to high generalization accuracy on unseen material.

The first action of the WPS method is to divide the data set D into a 80% training subset and a remaining 20% test subset. Let n be the number of labeled examples in the training subset. A parabolic sequence of d data set sizes is created from this training subset by using a factor $f = \sqrt[d]{n}$. We set the default number of steps $d = 20$. Starting with a seed data set of one example, a parabolically increasing sequence of $i = \{1 \ldots d\}$ data set sizes $size_i$ is created by letting $size_1 = 1$ and for every $i > 1$, $size_i = size_{i-1} * f$. We then limit the generated list of 20 sizes down to a list containing only the data sets with more than 500 examples. We also include the 500-example data set itself as the first set. This leaves a clipped pseudo-parabolic series. For each of the training sets, an accompanying test set is created by taking, from the tail of the 20% test subset, a set that has 20% of the size of its corresponding training set.

The WPS procedure is an iterative procedure over the clipped list of data set sizes. The procedure operates on a pool of settings, S, where one setting is a unique combination of algorithmic parameter values of IB1. At the outset, S contains all possible combinations of values of IB1's

7.1. WRAPPED PROGRESSIVE SAMPLING

parameters. We refer to them as $s_1 \ldots s_c$, c being the total number of possible combinations.

The first step of WPS is to perform experiments with all settings in $s_1 \ldots s_c$. Each of these experiments involves training IB1 on the first training set (500 examples) and testing the learned model on the first test set (100 examples), and measuring IB1's test accuracy, viz. the percentage of correctly classified test examples. This produces a list of accuracies, $acc(s_1) \ldots acc(s_c)$. As the second step, badly-performing settings from the current set are removed on grounds of their low score. This selection is performed with some care, since it is unknown whether a setting that is currently performing badly, would perform better than other settings when trained on more examples. WPS, therefore, does not simply sort the list of accuracies and cut away the lower-performing part of some predefined fraction. Rather, it attempts to estimate at each step the subset of accuracies that stands out as the best performing group, whichever portion of the total set of accuracies that is. To this end, a simple linear histogram is computed on all accuracies, dividing them in ten equally-wide bins, $b_1 \ldots b_{10}$ (the notation for the size of a bin, the number of accuracies in the bin, is $|b_i|$).

Without assuming any distribution over the bins, WPS enacts the following procedure to determine which settings are to be selected for the next step. First, the bin with the highest accuracies is taken as the first selected bin. Subsequently, every preceding bin is also selected that represents an equal number of settings or more than its subsequent bin, $|b_i| \geq |b_{i+1}|$. This is determined in a loop that halts as soon as $|b_i| < |b_{i+1}|$. The motivation behind this histogram-based selection is that it avoids the assumption that the accuracies are normally distributed – which is often not the case. A normal distribution assumes one peak, while the distribution of outcomes often contains multiple peaks. The current method adapts itself to the actual distribution by taking half of the best-performing peak (including the top of the peak), whether this is a perfect 50% right-hand side of a normal distribution or not.

Next, all non-selected settings are deleted from S, and the next step is initiated. This involves discarding the current training set and test set, and replacing them by their next-step progressively sampled versions. On this bigger-sized training and test set combo, all settings in S are tested through experiments, a histogram is computed on the outcomes, etcetera.

The process is iterated until either one of these stop conditions is met: (1) After the most recent setting selection, only one setting is left. Even if more training set sizes are available, these are not used, and the search halts, returning the one selected setting. Or, (2) after the last setting selection on the basis of experiments with the largest training and test set sizes, several settings are still selected. First, it is checked whether IB1's default setting is among them. If it is, this default setting is returned. If not, a random selection is made among the selected settings, and the randomly chosen setting is returned.

We customized WPS to IB1 by specifying five of the latter's parameters, and identifying a number of values for each of these parameters. The following list enumerates all tested settings. Due to the constraints mentioned in items 4 and 5, the total number of combinations of settings tested in the first round of the WPS procedure totals to $5 + (5 \times 2 \times 2) + (9 \times 5 \times 4) + (9 \times 5 \times 4 \times 2 \times 2) = 925$ settings. See chapter 3 for more details on the parameters.

1. -k determines how many groups of equidistant nearest neighbors are used to classify a new example. We vary among values $1, 3, 5, 7, 9, 11, 13, 15, 19, 25, 35$. Default is 1.
2. -w determines the employed feature weight in the similarity metric. The five options are no weighting, information gain, gain ratio, χ^2, or shared variance. Default is gain ratio.
3. -m determines the basic type of similarity metric. The choice is between overlap (default), MVDM, or Jeffrey divergence.
4. Only with -k set to a value larger than one, is distance weighting possible (-d). Options are to do no weighting (default), or perform inverse-linear weighting, inverse weighting, or weighting through exponential decay with $\alpha = 1$.
5. Only with -m with MVDM or Jeffrey divergence is it possible to back-off to the overlap metric through the -L parameter. We vary between 1, 2. Default is 1.

7.1.2 Getting started with wrapped progressive sampling

Wrapped progressive sampling is built into the command-line tool **paramsearch**. For data sets larger than 1,000 examples it performs the search procedure exactly as described above; for smaller data sets, it reverts to simple wrapping through internal n-fold cross-validation on the training material. Assuming, for now, a labeled training set with over 1,000 examples, paramsearch works as follows.

7.1. WRAPPED PROGRESSIVE SAMPLING

Using the full `gplural.data` data set representing our German plural task (cf. section 3.1), the first command is:

```
% paramsearch ib1 gplural.train
```

After establishing some training set statistics, paramsearch starts the first pseudo-exhaustive WPS step with a fixed training subset of 500 examples, and a test subset of 100 examples:

```
paramsearch v 1.0

gplural.data has
            25168 instances
                7 features (lowest value frequency 1)
                8 classes
optimizing algorithmic parameters of ib1
running wrapped progressive sampling parameter search
multiplication factor for steps: 1.641332
starting wrapped progressive sampling with first
pseudo-exhaustive round, stepsize 500
925 settings in current selection
computing density in 10% intervals between
lowest  44.00 and highest  86.00
density block 0 ( 44.00 -  48.20):    25
density block 1 ( 48.20 -  52.40):     3
density block 2 ( 52.40 -  56.60):     8
density block 3 ( 56.60 -  60.80):     7
density block 4 ( 60.80 -  65.00):    12
density block 5 ( 65.00 -  69.20):    18
density block 6 ( 69.20 -  73.40):    22
density block 7 ( 73.40 -  77.60):    79
density block 8 ( 77.60 -  81.80):   444
density block 9 ( 81.80 -  86.00):   307
decreasing density point before block 0
keeping the top 751 settings with accuracy  77.60 and up
```

After the 925 experiments have been completed, the distribution of accuracies of the 925 tested settings is visualized in ten lines. Each line represents one equal-width bar of the histogram, and mentions at its end the number of settings in the bar. For example, in the tenth bar, containing accuracies between 81.80% and 86.00%, 307 settings can be found.

To determine the settings to be tested in the next round, paramsearch selects all settings in the ninth and tenth bar (the ninth bar contains more settings than the tenth, but the eighth bar marks a decrease in the number of settings). This joint subset of 751 settings then proceeds to the second step.

With the German plural data it takes eight steps to eventually arrive at a single best setting, which is written to a file that has the same name as the training set, with the extension .ib1.bestsetting. Paramsearch also displays the setting on screen, in the format which is used in TIMBL output files, along with the accuracy on test data that this setting attained in the last step of the WPS procedure.

```
best setting found:
92.825104 gplural.data.test.IB1.J.L1.sv.k11.ID.%
wrapped progressive sampling process finished in 441 seconds
```

7.1.3 Wrapped progressive sampling results

To exemplify the effect of WPS we apply it to the six tasks which have been under investigation in the previous chapter as well: three non-sequential tasks, GPLURAL, DIMIN, and PP, and three sequential tasks, MORPH, CHUNK, and NER. The non-sequential tasks are evaluated on accuracy (% of correctly classified test instances), while the sequential tasks are evaluated on accuracy, precision, recall, and F-score of predicted analyses and structures:

1. GPLURAL, German plural formation
2. DIMIN, Dutch diminutive formation
3. PP, English prepositional phrase attachment
4. MORPH, Dutch morphological analysis
5. CHUNK, English base phrase chunking
6. NER, English named entity recognition

Tables 7.1 and 7.2 list the generalization performances in terms of accuracy for GPLURAL, DIMIN, and PP (Table 7.1), and overall precision, recall, and F-score, as measured by the publicly available evaluation scripts for CHUNK (Tjong Kim Sang & Buchholz, 2000) and NER (Tjong Kim Sang & De Meulder, 2003) and with analogous evaluations performed for MORPH (cf. section 4.2). The right-hand column of the tables lists the percentage of error reduction (or error increase, denoted by a negative number) produced by WPS. For PP and MORPH we observe slight increases of error by 3% (0.6 points of accuracy) and 1% (0.3 points of F-score), respectively. For the other tasks, modest to considerable reductions are observed, between 4%

7.1. WRAPPED PROGRESSIVE SAMPLING

Task	Generalization performance (accuracy) Without WPS	With WPS	Error reduction (%)
GPLURAL	94.6	94.8	4
DIMIN	96.7	97.5	24
PP	81.3	80.7	-3

Table 7.1: Comparison of generalization performances in terms of accuracy (Acc) on benchmark NLP test sets, without and with parameter optimization by wrapped progressive sampling. The right-hand column displays the percentage of error reduction.

Task	Generalization performance						Error reduction (%)
	Without WPS			With WPS			
	Pre.	Rec.	F	Pre.	Rec.	F	
MORPH	71.9	68.9	70.4	71.8	68.4	70.1	-1
CHUNK	89.4	91.3	90.3	91.1	92.6	91.9	16
NER	64.8	69.0	66.8	76.6	77.8	77.2	31

Table 7.2: Comparison of generalization performances in terms of precision (Pre.), recall (Rec.), and F-score (F) on benchmark NLP test sets, without and with parameter optimization by wrapped progressive sampling. The right-hand column displays the percentage of error reduction in F-score.

(0.2 points of F-score) for GPLURAL, to a considerable 31% (10.4 points of F-score) for NER.

The settings that are estimated to be optimal by WPS are listed in Table 7.3. What most distinguishes these settings from IB1's default setting is that k is never set to 1. Also, the Overlap metric is only selected for GPLURAL. Distance weighting is used with all tasks except NER.

The results presented here suggest a generally harmless and mostly positive effect of performing WPS. It could very well act as a useful addition to experiments with memory-based learning. In the remainder of this chapter we continue to experiment on the four sequence-based tasks. In these experiments, we apply WPS to every training set, and use the produced setting for the final test.

	Parameter				
Task	-k	-w	-m	-d	-L
GPLURAL	3	IG	O	IL	–
DIMIN	11	X2	J	IL	1
PP	7	GR	J	ID	2
MORPH	11	SV	M	ID	2
CHUNK	25	GR	M	IL	2
NER	3	GR	M	Z	1

Table 7.3: Estimated-optimal parameter settings found through wrapped progressive sampling corresponding with the results in Tables 7.1 and 7.2.

7.2 Optimizing output sequences

Many tasks in natural language processing are sequence tasks, due to the obvious sequential nature of words as sequences of phonemes or letters, and sentences and spoken utterances as sequences of words. However, many machine learning methods, including memory-based learning, do not typically learn these tasks by learning to map input sequences to output sequences. Rather, the standard approach (as also exemplified in earlier chapters) is to encode a sequence processing task by windowing, in which input sequences are mapped to one output symbol, which is typically a symbol associated with one of the input symbols, for example the middle one in the window. There are but a few current machine-learning algorithms that map sequences to sequences though a monolithic model; two current examples are maximum-entropy markov models (McCallum et al., 2000) and conditional random fields (Lafferty et al., 2001). Other approaches exist that enhance a simple (e.g., windowing-based) model with internal feedback loops, so that the model can learn from its own previous decisions (for example, the MBT system described in chapter 5, and in recurrent neural networks Elman, 1990; Sun & Giles, 2001).

Figure 7.1 displays this simplest version of the windowing process; fixed-width subsequences of input symbols are coupled to one output symbol. To ignore that the output forms a sequence is a problematic restriction, since it allows the classifier to produce invalid or impossible output sequences. The problem is a double one, triggering two different solutions:

7.2. OPTIMIZING OUTPUT SEQUENCES

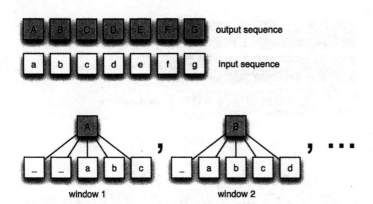

Figure 7.1: Standard windowing process. Sequences of input symbols and output symbols are converted into windows of fixed-width input symbols each associated with one output symbol.

1. Classifications, producing an output symbol, are based on sequences (windows) of input symbols only. Due to this, a classifier can make two neighboring classifications in a sequence that are not compatible with each other, since it has no information about the other decision.

2. The very fact that classifications produce single output symbols is a restriction that is not intrinsic – the task may well be rephrased so that each input window is mapped to a sequence of output symbols. This directly prevents the classifier from ever predicting an invalid output sequence, since it will always produce sequences it has learned from training material.

In 7.2.1 we present a solution to the first problem that does not involve a classifier-internal feedback loop nor is tied to directional processing. Instead the solution is based on the idea of classifier stacking, in which a second classifier corrects the output of a near-sighted first classifier. The second solution, to predict sequences of output symbols directly, is described in subsection 7.2.2. The two approaches are tied together in subsection 7.2.3, in which we show that the two solutions are partly complementary; they prevent different errors.

7.2.1 Stacking

Stacking, a term popularized by Wolpert (1992) in an artificial neural network context, refers to a class of meta-learning systems that learn to

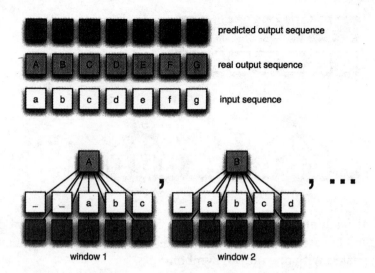

Figure 7.2: The windowing process after a first-stage classifier has produced a predicted output sequence. Sequences of input symbols, predicted output symbols, and real output symbols are converted into windows of fixed-width input symbols and predicted output symbols each associated with one output symbol.

correct errors made by lower-level classifiers. Stacking applied to memory-based learning was introduced in (Veenstra, 1998), and later used in (Hendrickx & Van den Bosch, 2003; Van den Bosch et al., 2004). We implement stacking by adding a windowed sequence of previous and subsequent output class labels to the original input features, and providing these enriched examples as training material to a second-stage classifier. Figure 7.2 illustrates the procedure. Given the (possibly erroneous) output of a first classifier on an input sequence, a certain window of class symbols from that predicted sequence is copied to the input, to act as predictive features for the real class label.

To generate the output of a first-stage classifier, two options are open. We name these options **perfect** and **adaptive**, to use the terms introduced in (Van den Bosch, 1997). They differ in the way they create training material for the second-stage classifier:

Perfect – the training material is created straight from the training material of the first-stage classifier, by windowing over the real class sequences. In so doing, the class label of each window is excluded from the input window, since it is always the same as the class to be predicted.

7.2. OPTIMIZING OUTPUT SEQUENCES

	\multicolumn{9}{c}{Generalization performance (%)}									
	No stacking			Perfect stacking				Adaptive stacking		
	Pre.	Rec.	F	Pre.	Rec.	F	Red.	Pre.	Rec.	F	Red.
---	---	---	---	---	---	---	---	---	---	---	---
MORPH	71.8	68.4	70.1	67.4	69.0	68.1	-7	74.2	72.2	73.2	10
CHUNK	91.1	92.6	91.9	91.3	92.7	92.0	1	92.1	93.1	92.6	11
NER	76.6	77.8	77.2	78.0	78.5	78.3	1	78.8	78.9	78.9	7

Table 7.4: Comparison of generalization performances in terms of precision (Pre.), recall (Rec.), and F-score (F) on three benchmark test sets, without stacking, and with perfect and adaptive stacking. The relative error reduction of the F-score of the two latter variants as compared to the "No stacking" F-score is given in de "Red." columns.

In training, this focus feature would receive an unrealistically high weight, especially considering that in testing this feature would contain errors. To assign a very high weight to a feature that may contain an erroneous value does not seem a good idea.

Adaptive – the training material is created indirectly by running an internal 10-fold cross-validation experiment on the first-stage training set, concatenating the predicted output class labels on all of the ten test partitions, and converting this output to class windows. In contrast to the perfect variant, we do include the focus class feature in the copied class label window. The adaptive approach can in principle learn from recurring classification errors in the input, and predict the correct class in case an error recurs.

Table 7.4 lists the comparative results on the MORPH, CHUNK, and NER tasks introduced earlier. They show that stacking generally works for these three tasks, but that the adaptive stacking variant produces higher relative gains than the perfect variant. There is a loss of about 2 points of F-score (an error increase of 7%) in the perfect variant of stacking applied to MORPH, on which the adaptive variant attains an error reduction in F-score of 10%. On CHUNK and NER the adaptive variant produces higher gains than the perfect variant as well; in terms of error reduction in F-score as compared to the situation without stacking, the gains are 11% for CHUNK and 7% for NER. There appears to be more useful information in training data derived from cross-validated output with errors, than in training data with error-free material.

Overall, the error reductions produced by stacking – up to 11% – are quite positive. From the results presented here it appears there is almost always something to be learned from taking into account a context of class labels of a first-stage classifier, even if the classifier is partly erroneous; particularly the adaptive variant of stacking attains the best gains overall.

7.2.2 Predicting class n-grams

A single memory-based classifier produces one class label at a time, but there is no intrinsic bound to what is packed into this class label. We exemplified in section 4.2 how operations can be packed into class labels and in section 5.2.2 that tagging and chunking decisions can be combined into one class system. Here, we show how class labels can span over n-grams of neighboring class labels. Although simple and appealing, the lurking disadvantage of this idea is that the amount of class labels increases explosively when moving from single class labels to wider n-grams. The CHUNK data, for example, have 22 classes ("IOB" codes associated with chunk types); in the same training set, 846 different trigrams of these 22 classes occur. Although this is far less than the theoretical maximum of $22^3 = 10,648$, it is still a sizeable number of classes, with fewer examples per trigram class in the training set than the original 22. MORPH, in turn, has 3,831 unigram classes, and a staggering 14,795 trigram classes. With NER, the situation is better; the eight original classes combine to 138 occurring trigrams.

Memory-based classification is insensitive to the actual number of different classes in the data, but the question is whether using n-grams of classes pushes the data sparseness to such levels that components of the similarity function are negatively affected.

Figure 7.3 illustrates the procedure by which windows are created with, as an example, class trigrams. Each windowed instance maps to a class label that incorporates three atomic class labels, namely the focus class label that was the original unigram label, plus its immediate left and right neighboring class labels.

While creating instances this way is trivial, it is not entirely trivial how the output of overlapping class trigrams recombines into a normal string of class sequences. When the example illustrated in Figure 7.3 is followed, each single class label in the output sequence is effectively predicted three times; first, as the right label of a trigram, next as the middle label, and finally as the left label of a trigram. Although it could be possible to avoid overlaps and classify only every n words (where n is the n-gram width), there is an interesting property of overlapping class label n-grams: namely, it is possible to apply voting to them.

7.2. OPTIMIZING OUTPUT SEQUENCES

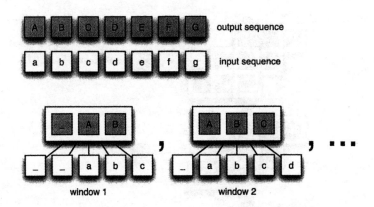

Figure 7.3: Windowing process with n-grams of class symbols. Sequences of input symbols and output symbols are converted into windows of fixed-width input symbols each associated with, in this example, trigrams of output symbols.

To pursue our example of trigram classes, the following voting procedure can be followed to decide about the resulting unigram class label sequence – referring to Figure 7.4 for visualization:

1. When all three votes are unanimous, their common class label is returned (Figure 7.4, top);
2. When two out of three vote for a common class label, this class label is returned (Figure 7.4, middle);
3. When all three votes disagree, the class label is returned of which the nearest neighbor is closest (Figure 7.4, bottom).

Clearly this scheme is one out of many possible schemes, using variants of voting as well as variants of n (and having multiple classifiers with different n, so that some back-off procedure could be followed). For now we use this procedure with trigrams as an example. To measure its effect we apply it again to the sequence tasks MORPH, CHUNK, and NER. The results of this experiment, where in each case WPS was used to find optimal algorithmic parameters, are listed in Table 7.5. For CHUNK and NER, the two data sets with relatively few class trigrams, the effect is rather positive; an error reduction of 12% with CHUNK, and 13% with NER. The task with the largest number of class trigrams, MORPH, is still learned with 6% error reduction.

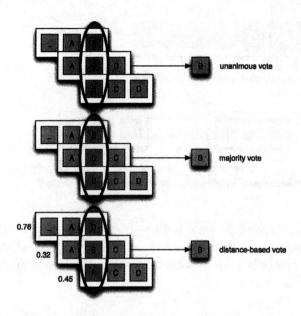

Figure 7.4: Three possible outcomes in voting over class trigrams: unanimous voting, majority voting, or distance-based voting.

	Generalization performance						
	No class n-grams			Class n-grams			Error
	Pre.	Rec.	F	Pre.	Rec.	F	reduction (%)
MORPH	71.8	68.4	70.1	73.2	70.5	71.8	6
CHUNK	91.1	92.6	91.9	92.8	92.9	92.8	12
NER	76.6	77.8	77.2	80.8	79.6	80.2	13

Table 7.5: Comparison of generalization performances in terms of precision (Pre.), recall (Rec.), and F-score (F) on three benchmark test sets without and with class n-grams. The right-hand column displays the error reduction in F-score by the class n-grams method over the other method.

7.2.3 Combining stacking and class n-grams

Stacking and class n-grams can be combined. One possible straightforward combination is that of a first-stage classifier that predicts n-grams, and a second-stage stacked classifier that also predicts n-grams (we use the adaptive variant, since it produced the best results), while including a window of first-stage n-gram class labels in the input, as illustrated in Figure 7.5.

7.2. OPTIMIZING OUTPUT SEQUENCES

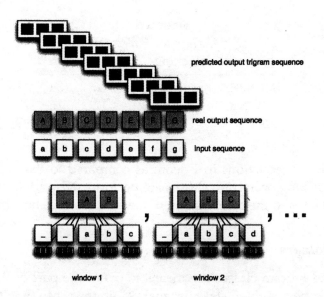

Figure 7.5: Windowing process with n-grams of class symbols after a first stage classifier has produced a sequence of n-grams of class symbols. Sequences of input symbols and output n-grams (trigrams) are converted into windows of fixed-width input symbols and output trigrams, each associated with one trigram of output symbols.

	Generalization performance (%)								
	Adaptive stacking			n-gram classes			Combination		
	Pre.	Rec.	F	Pre.	Rec.	F	Pre.	Rec.	F
MORPH	74.2	72.2	73.2	73.2	70.5	71.8	80.8	79.9	80.3
CHUNK	92.1	93.1	92.6	92.8	92.9	92.8	93.2	93.0	93.1
NER	78.8	78.9	78.9	80.8	79.6	80.2	81.6	79.6	80.6

Table 7.6: Comparison of generalization performances in terms of precision (Pre.), recall (Rec.), and F-score (F) on three benchmark test sets, with adaptive stacking, n-gram classes, and the combination of the two.

Table 7.6 compares the results of adaptive stacking and n-gram classes with those of the combination. As can be seen, the combination produces even better results than both stacking and n-gram classes for the MORPH, CHUNK, and NER tasks; compared to the original setting without stacking or n-gram classes, the error reduction is no less than 34% for MORPH, 16% for CHUNK, and 15% for NER.

	Error reduction (%)		
	Adaptive stacking	n-gram classes	Both
MORPH	10	6	34
CHUNK	11	12	16
NER	7	13	15

Table 7.7: Error reductions in F-score as compared to the experiment in which no stacking or n-gram classes are performed, by adaptive stacking, n-gram classes, and the combination of the two, on three benchmark tasks.

7.2.4 Summary

Stacking and n-gram classes are means to overcome part of the mistakes that a simple near-sighted classifier makes – in the sense that it is blind to its own decisions elsewhere in the output symbol sequence – in processing sequential NLP tasks. We showed that with IB1 both methods tend to lead to better performances on a range of tasks. Stacking appears to work best in its adaptive variant, in which a second-stage classifier is trained on examples that contain the partly erroneous output of a first-stage classifier; we observed error reductions in F-score of up to 11%. n-gram classification proved to be at least as useful; we noted error reductions of up to 13%. On top of that, the combination of the two is able to reduce errors down by 15% to 34% as compared to the situation without stacking or n-gram classes; the two methods solve partly different near-sightedness problems.

The error reductions in F-score of the four tasks are reproduced in Table 7.7. A surprisingly positive result is the error reduction on the MORPH task. The combination reduces more error, 34% (10.2 points of F-score) than the sum of the effects of the two methods in isolation (10% plus 6%). This suggests that the two optimizations strengthen each other – an exciting result.

7.3 Conclusion

Handling sequences in NLP is commonly seen as the area for generative models such as Hidden Markov Models (HMMs) and stochastic grammars (Manning & Schütze, 1999). Improvements such as Conditional Random Fields (Lafferty et al., 2001) are motivated by limitations of the former generative models, which have trouble with interacting features or long-distance dependencies. It appears that symbolic classifiers that do not optimize some likelihood over output sequences have no place in sequence tasks. It is, indeed, easy to give examples of strange errors made by such classifiers that display their weakness: each prediction is made in total oblivion of other predictions made in the sequence.

However, in this chapter we provided two complementary examples of simple manipulations of the feature space and the class space that allow the very same near-sighted classifiers to repair some of their stupid errors, without the incorporation of higher-level likelihood optimization or solution reranking. The first variation involves stacking, an architecture which connects a second-stage classifier to a near-sighted first-stage classifier. The second classifier is trained both on the original features and on a sequenced output of the first-stage classifier's predictions. On three tasks we attain reasonable error reductions of 7 to 10%.

The second variant uses a single classifier, but changes the class space by letting the classifier predict n-grams of classes. This way the classifier is forced to predict syntactically valid n-grams. We only tested class trigrams and introduced a voting scheme that uses all overlapping trigrams for one sequence, converting the trigrams to a voted series of single predictions, and attained reasonable error reductions of 6 to 13%. Interestingly, when the two variants are combined the error reductions increase (to between 15 to 34%). The two variants appear to strengthen rather than mute each other; they seem to repair different errors. Further research into the exact types of errors repaired by the two methods is needed.

The actual comparison between these variants of memory-based learning and generative models still needs to be made. Also, the variants are not specific to memory-based learning; they work with any classifier that suffers from the same near-sightedness problem. This calls for large-scale comparative studies, which in our opinion the general field of data-driven NLP needs to deepen its understanding of the learnability of sequential NLP tasks such as parsing.

7.4 Further reading

The overview of memory-based language processing work as presented in this book is not complete – as much as we have attempted to identify central topics, application areas, open issues, and extensions, as well as given pointers to further reading, there is much work that we have not addressed so far. To redress the balance somewhat, we take the opportunity in this section to provide pointers to memory-based language processing research in a number of NLP applications and in cognitive linguistics.

Word sense disambiguation is a task that involves selecting the contextually correct sense of a word given features representing properties of the word to be disambiguated and its context. Work in our own groups has focused on the role of algorithm parameter optimization (see section 7.1) in a word expert approach to the task, with very good results in the SENSEVAL workshops[1] (Veenstra et al., 2000; Hoste et al., 2002; Decadt et al., 2004). Other groups have used TIMBL for this task as well, for example for the integration of multiple sources of information (Stevenson & Wilks, 1999). An alternative instance-based approach to word sense disambiguation is described in Mihalcea (2002). A special case is the ambiguous English word *it* which has been the focus of a disambiguation study using TIMBL by Evans (2001). Related lexical disambiguation tasks in which TIMBL is used are noun countability, explored in a study by Baldwin and Bond (2003), and noun animacy (Orăsan & Evans, 2001).

Anaphora resolution has been attempted with memory-based methods as well: Preiss (2002) used TIMBL to weigh the relevance of features, whereas Hoste (2005) provides a full memory-based solution to the task, and compares it with eager learning methods. Mitkov et al. (2002) use classification with TIMBL to disambiguate co-reference relations of *it* (extending the aforementioned study on the disambiguation of *it*) in a knowledge-poor pronoun resolution system.

In the context of *(spoken) dialog systems*, MBLP has been successfully applied to tasks ranging from the interpretation of spoken input (word graphs) to miscommunication detection and dialog act classification (Van den Bosch et al., 2001; Lendvai et al., 2002; Lendvai et al., 2003a). The detection of disfluencies in spoken language has been investigated in Lendvai et al. (2003b). TIMBL is employed to assign dialog acts to elliptic *sluice* phenomena (Fernández et al., 2004).

In *speech generation*, TIMBL has been applied to the problem of prosody generation for Dutch, by reducing the task of prosody generation to the dual classification task of predicting the appropriate location of phrase boundaries and sentence accents in a windowing approach (Marsi et al., 2003).

At the level of *text generation*, the order of prenominal adjectives has

---

[1] SENSEVAL provides data and evaluates output of word sense disambiguation systems on these data. See http://www.senseval.org/

7.4. FURTHER READING

been predicted using TIMBL (Malouf, 2000), as well as the generation of determiners (Minnen et al., 2000). Sentence generation using memory-based learning has been investigated by Varges and Mellish (2001).

At the level of *text mining* and information system applications (information extraction, document classification, etc.), TIMBL has been used in information extraction (Zavrel et al., 2000; Zavrel & Daelemans, 2003), spam filtering (Androutsopoulos et al., 2000), and generic text classification (Spitters, 2000). As a special case of information extraction, memory-based models of named-entity recognition (identifying and labeling proper names of persons, locations, organizations, and other entities) have been proposed in Buchholz and Van den Bosch (2000); Hendrickx and Van den Bosch (2003); De Meulder and Daelemans (2003).

Although we have sometimes used linguistically relevant data sets (such as the German plural data set introduced in chapter 3), the focus of this book has been on algorithms and applications in computational linguistics. Nevertheless, memory-based language processing has also been suggested as a psychologically and linguistically relevant account of human language behavior. For several tasks, human behavior in acquisition and processing has been successfully matched to the output of MBLP algorithms. A good illustration of this approach is work on the assignment of stress to words in Dutch and other languages, which is traditionally described in a principles and parameters or optimality theory nativist framework: variation in stress systems is described by means of a number of principles (innate rules or constraints) that are "tuned" in acquisition by parameter setting or constraint ordering to a specific language. Research using memory-based language processing shows that data-oriented methods presupposing only minimal innate structure are able to learn both the *core* and the *periphery* of language-specific stress systems, and that the output and errors generated by these analogical systems also match the available psycholinguistic processing and acquisition data (Daelemans et al., 1994; Durieux & Gillis, 2000; Gillis et al., 2000). More recent work by psycholinguists and linguists using TIMBL similarly shows the descriptive adequacy of models based on memory and analogical reasoning for problems in morphology (Krott et al., 2001; Eddington, 2003).

Bibliography

Aamodt, A., & Plaza, E. (1994). Case-based reasoning: Foundational issues, methodological variations, and system approaches. *AI Communications, 7*, 39–59.

Abney, S. (1991). Parsing by chunks. In *Principle-based parsing*, 257–278. Dordrecht: Kluwer Academic Publishers.

Abney, S., Schapire, R., & Singer, Y. (1999). Boosting Applied to Tagging and PP Attachment. *Proceedings of the 1999 Joint SIGDAT Conference on Empirical Methods in Natural Language Processing and Very Large Corpora*, 38–45.

Aha, D. W. (Ed.). (1997). *Lazy learning*. Dordrecht: Kluwer Academic Publishers.

Aha, D. W., Kibler, D., & Albert, M. (1991). Instance-based learning algorithms. *Machine Learning, 6*, 37–66.

Allen, J. (1995). *Natural language understanding*. Redwood City, CA: The Benjamin/Cummings Publishing Company. Second edition.

Allen, J., Hunnicutt, M. S., & Klatt, D. (1987). *From text to speech: The MITalk system*. Cambridge, England: Cambridge University Press.

Androutsopoulos, I., Paliouras, G., Karkaletsis, V., Sakkis, G., Spyropoulos, C. D., & Stamatopoulos, P. (2000). Learning to filter spam e-mail: A comparison of a Naive Bayesian and a memory-based approach. *Proceedings of the "Machine Learning and Textual Information Access" Workshop of the 4th European Conference on Principles and Practice of Knowledge Discovery in Databases*.

Argamon, S., Dagan, I., & Krymolowski, Y. (1999). A memory-based approach to learning shallow natural language patterns. *Journal of Experimental and Theoretical Artificial Intelligence, special issue on memory-based learning, 10*, 1–22.

Atkeson, C., Moore, A., & Schaal, S. (1997). Locally weighted learning. *Artificial Intelligence Review, 11*, 11–73.

Baayen, R. H., Piepenbrock, R., & van Rijn, H. (1993). *The CELEX lexical data base on CD-ROM*. Philadelphia, PA: Linguistic Data Consortium.

Bailey, T., & Jain, A. K. (1978). A note on distance-weighted k-nearest neighbor rules. *IEEE Transactions on Systems, Man, and Cybernetics, SMC-8*, 311–313.

Baker, C., Fillmore, C., & Lowe, J. (1998). The Berkeley FrameNet project. *Proceedings of the COLING-ACL*, 86–90. Montreal, Canada.

Baldwin, T., & Bond, F. (2003). A plethora of methods for learning English countability. *Proceedings of the 2003 Conference on Empirical Methods in Natural Language Processing*, 73–80. New Brunswick, NJ: ACL.

Banko, M., & Brill, E. (2001). Scaling to very very large corpora for natural language disambiguation. *Proceedings of the 39th Annual Meeting of the Association for Computational Linguistics*, 26–33. Association for Computational Linguistics.

Barlow, M., & Kemmer, S. (2000). *Usage-based models of language*. Stanford: CSLI Publications.

Bentley, J. L., & Friedman, J. H. (1979). Data structures for range searching. *ACM Computing Surveys, 11*, 397–409.

Bloomfield, L. (1933). *Language*. New York, NY: Holt, Rinehard and Winston.

Bod, R. (1995). *Enriching linguistics with statistics: Performance models of natural language*. Doctoral dissertation, ILLC, Universiteit van Amsterdam, Amsterdam, The Netherlands.

Bod, R. (1998). *Beyond grammar: An experience-based theory of language*. CSLI Publications. Cambridge University Press.

Boland, J. E., & Boehm-Jernigan, H. (1998). Lexical constraints and prepositional phrase attachment. *Journal of Memory and Language, 39*, 684–719.

Booij, G. (2001). *The morphology of Dutch*. Oxford, UK: Oxford University Press.

Brants, T. (2000). TnT – a statistical part-of-speech tagger. *Proceedings of the 6th Applied NLP Conference, ANLP-2000, April 29 – May 3, 2000, Seattle, WA*.

Breiman, L., Friedman, J., Ohlsen, R., & Stone, C. (1984). *Classification and regression trees*. Belmont, CA: Wadsworth International Group.

Brighton, H., & Mellish, C. (2002). Advances in instance selection for instance-based learning algorithms. *Data Mining and Knowledge Engineering, 6*, 153–172.

Brill, E. (1992). A simple rule-based part of speech tagger. *Proceedings of the DARPA Workshop on Speech and Natural Language*.

Brill, E. (1994). Some advances in transformation-based part-of-speech tagging. *Proceedings AAAI '94*.

Brill, E. (1995). Transformation-based error-driven learning and natural language processing: A case study in part of speech tagging. *Computational Linguistics, 21*, 543–565.

Brill, E., & Mooney, R. (Eds.). (1998a). *The AI magazine: Special issue on empirical natural language processing*, vol. 18. AAAI.

Brill, E., & Mooney, R. J. (1998b). An overview of empirical natural language processing. *The AI Magazine, 18,* 13–24.

Brill, E., & Resnik, P. (1994). A rule-based approach to prepositional phrase attachment disambiguation. *Proceedings of 15th annual conference on Computational Linguistics.*

Buchholz, S. (1998). Distinguishing complements from adjuncts using memory-based learning. *Proceedings of the ESSLLI-98 Workshop on Automated Acquisition of Syntax and Parsing, Saarbrücken, Germany.*

Buchholz, S. (2002). *Memory-based grammatical relation finding.* PhD thesis, University of Tilburg.

Buchholz, S., & Daelemans, W. (2001). Complex Answers: A Case Study using a WWW Question Answering System. *Journal for Natural Language Engineering.*

Buchholz, S., & Van den Bosch, A. (2000). Integrating seed names and n-grams for a named entity list and classifier. *Proceedings of the Second International Conference on Language Resources and Evaluation,* 1215–1221. Athens, Greece.

Buchholz, S., Veenstra, J., & Daelemans, W. (1999). Cascaded grammatical relation assignment. *EMNLP-VLC'99, the Joint SIGDAT Conference on Empirical Methods in Natural Language Processing and Very Large Corpora.*

Busser, G. (1998). Treetalk-d: a machine learning approach to Dutch word pronunciation. *Proceedings of the Text, Speech, and Dialogue Conference,* 3–8.

Canisius, S., & Van den Bosch, A. (2004). A memory-based shallow parser for spoken Dutch. In B. Decadt, G. De Pauw and V. Hoste (Eds.), *Selected papers from the thirteenth computational linguistics in the Netherlands meeting,* 31–45. University of Antwerp.

Cardie, C. (1996). Automatic feature set selection for case-based learning of linguistic knowledge. *Proceedings of the Conference on Empirical Methods in NLP.*

Cardie, C., & Mooney, R. (Eds.). (1999). *Machine learning: Special issue on machine learning and natural language,* vol. 34. Kluwer Academic Publishers.

Carl, M., & Way, A. (2003). *Recent advances in example-based machine translation,* vol. 21 of *Text, Speech and Language Technology.* Dordrecht: Kluwer Academic Publishers.

Charniak, E. (1993). *Statistical Language Learning.* Cambridge, MA: The MIT Press.

Chomsky, N., & Halle, M. (1968). *The sound pattern of English.* New York, NY: Harper & Row.

Church, K., & Mercer, R. L. (1993). Introduction to the Special Issue on Computational Linguistics Using Large Corpora. *Computational Linguistics, 19,* 1–24.

BIBLIOGRAPHY

Clahsen, H. (1999). Lexical entries and rules of language: A multidisciplinary study of German inflection. *Behavioral and Brain Sciences*, 22, 991–1060.

Clark, A. (2002). Memory-based learning of morphology with stochastic transducers. *Proceedings of the 40th Meeting of the Association for Computational Linguistics*, 513–520. New Brunswick, NJ: ACL.

Clark, P., & Boswell, R. (1991). Rule induction with CN2: Some recent improvements. *Proceedings of the Sixth European Working Session on Learning*, 151–163. Berlin: Springer Verlag.

Clark, P., & Niblett, T. (1989). The CN2 rule induction algorithm. *Machine Learning*, 3, 261–284.

Cohen, W. (1995). Fast effective rule induction. *Proceedings of the 12th International Conference on Machine Learning*, 115–123. Morgan Kaufmann.

Collins, M., & Brooks, J. (1995). Prepositional phrase attachment through a backed-off model. *Proceedings of the Third Workshop on Very Large Corpora*. Cambridge.

Cortes, C., & Vapnik, V. (1995). Support vector networks. *Machine Learning*, 20, 273–297.

Cost, S., & Salzberg, S. (1993). A weighted nearest neighbour algorithm for learning with symbolic features. *Machine Learning*, 10, 57–78.

Cover, T. M., & Hart, P. E. (1967). Nearest neighbor pattern classification. *Institute of Electrical and Electronics Engineers Transactions on Information Theory*, 13, 21–27.

Cristiani, N., & Shawe-Taylor, J. (2000). *An introduction to support vector machines*. Cambridge, UK: Cambridge University Press.

Croft, W., & Cruse, A. (2003). *Cognitive linguistics*. Cambridge Textbooks in Linguistics. Cambridge: Cambridge University Press.

Daelemans, W. (1995). Memory-based lexical acquisition and processing. In P. Steffens (Ed.), *Machine translation and the lexicon*, Lecture Notes in Artificial Intelligence, 85–98. Berlin: Springer-Verlag.

Daelemans, W. (1996). Experience-driven language acquisition and processing. In M. Van der Avoird and C. Corsius (Eds.), *Proceedings of the CLS opening academic year 1996-1997*, 83–95. Tilburg: CLS.

Daelemans, W. (2002). A comparison of analogical modeling to memory-based language processing. In R. Skousen, D. Lonsdale and D. Parkinson (Eds.), *Analogical modeling*. Amsterdam, The Netherlands: John Benjamins.

Daelemans, W., Berck, P., & Gillis, S. (1997a). Data mining as a method for linguistic analysis: Dutch diminutives. *Folia Linguistica*, XXXI, 57–75.

Daelemans, W., Gillis, S., & Durieux, G. (1994). The acquisition of stress: a data-oriented approach. *Computational Linguistics*, 20, 421–451.

Daelemans, W., & Hoste, V. (2002). Evaluation of machine learning methods for natural language processing tasks. *Proceedings of the Third International Conference on Language Resources and Evaluation*, 755–760. Las Palmas, Gran Canaria.

Daelemans, W., Höthker, A., & Tjong Kim Sang, E. (2004a). Automatic sentence simplification for subtitling in Dutch and English. *Proceedings of the 4th International Conference on Language Resources and Evaluation (LREC-04)*, 1045–1048.

Daelemans, W., & Van den Bosch, A. (1992). Generalisation performance of backpropagation learning on a syllabification task. *Proceedings of TWLT3: Connectionism and Natural Language Processing*, 27–37. Enschede.

Daelemans, W., & Van den Bosch, A. (1996). Language-independent data-oriented grapheme-to-phoneme conversion. In J. P. H. Van Santen, R. W. Sproat, J. P. Olive and J. Hirschberg (Eds.), *Progress in speech processing*, 77–89. Berlin: Springer-Verlag.

Daelemans, W., & Van den Bosch, A. (2001). Treetalk: Memory-based word phonemisation. In R. Damper (Ed.), *Data-driven techniques in speech synthesis*, 149–172. Dordrecht: Kluwer Academic Publishers.

Daelemans, W., Van den Bosch, A., & Weijters, A. (1997b). IGTree: using trees for compression and classification in lazy learning algorithms. *Artificial Intelligence Review*, 11, 407–423.

Daelemans, W., Van den Bosch, A., & Zavrel, J. (1999). Forgetting exceptions is harmful in language learning. *Machine Learning, Special issue on Natural Language Learning*, 34, 11–41.

Daelemans, W., Weijters, A., & Van den Bosch, A. (Eds.). (1997c). *Workshop notes of the ecml/mlnet familiarisation workshop on empirical learning of natural language processing tasks*. Prague, Czech Republic: University of Economics.

Daelemans, W., Zavrel, J., Berck, P., & Gillis, S. (1996). MBT: A memory-based part of speech tagger generator. *Proceedings of the Fourth Workshop on Very Large Corpora*, 14–27.

Daelemans, W., Zavrel, J., Van den Bosch, A., & Van der Sloot, K. (2003). *Mbt: Memory based tagger, version 2.0, reference guide* (Technical Report ILK 03-13). ILK Research Group, Tilburg University.

Daelemans, W., Zavrel, J., Van der Sloot, K., & Van den Bosch, A. (1998). *TiMBL: Tilburg Memory Based Learner, version 1.0, reference manual* (Technical Report ILK 98-03). ILK Research Group, Tilburg University.

Daelemans, W., Zavrel, J., Van der Sloot, K., & Van den Bosch, A. (2004b). *TiMBL: Tilburg Memory Based Learner, version 5.1.0, reference guide* (Technical Report ILK 04-02). ILK Research Group, Tilburg University.

Dagan, I., Lee, L., & Pereira, F. (1997). Similarity-based methods for word sense disambiguation. *Proceedings of the 35th Annual Meeting of the Association for Computational Linguistics and the 8th Annual Meeting of the European Chapter of the Association for Computational Linguistics*, 56–63.

Dale, R., Moisl, H., & Somers, H. (Eds.). (2000). *Handbook of natural language processing*. New York: Marcel Dekker Inc.

Damper, R. (1995). Self-learning and connectionist approaches to text-phoneme conversion. In J. Levy, D. Bairaktaris, J. Bullinaria and P. Cairns (Eds.), *Connectionist models of memory and language*, 117–144. London, UK: UCL Press.

Damper, R., & Eastmond, J. (1997). Pronunciation by analogy: impact of implementational choices on performance. *Language and Speech*, 40, 1–23.

Dasarathy, B. V. (1991). *Nearest neighbor (NN) norms: NN pattern classification techniques*. Los Alamitos, CA: IEEE Computer Society Press.

De Haas, W., & Trommelen, M. (1993). *Morfologisch handboek van het nederlands: Een overzicht van de woordvorming*. 's Gravenhage, The Netherlands: SDU.

De Meulder, F., & Daelemans, W. (2003). Memory-based named entity recognition using unannotated data. *Proceedings of CoNLL-2003*, 208–211. Edmonton, Canada.

De Pauw, G., Laureys, T., Daelemans, W., & Van hamme, H. (2004). A comparison of two different approaches to morphological analysis of Dutch. *Proceedings of the ACL 2004 Workshop on Current Themes in Computational Phonology and Morphology*, 62–69.

De Saussure, F. (1916). *Cours de linguistique générale*. Paris: Payot. Edited posthumously by C. Bally, A. Sechehaye, and A. Riedlinger. Citation page numbers and quotes are from the English translation by Wade Baskin, New York: McGraw-Hill Book Company, 1966.

Devijver, P. A., & Kittler, J. (1980). On the edited nearest neighbor rule. *Proceedings of the Fifth International Conference on Pattern Recognition*. The Institute of Electrical and Electronics Engineers.

Decadt, B., Hoste, V., Daelemans, W., & Van den Bosch, A. (2004). GAMBL, genetic algorithm optimization of memory-based WSD. *Proceedings of the Third International Workshop on the Evaluation of Systems for the Semantic Analysis of Text (Senseval-3)*, 108–112. New Brunswick, NJ: ACL.

Dedina, M. J., & Nusbaum, H. C. (1991). PRONOUNCE: a program for pronunciation by analogy. *Computer Speech and Language*, 5, 55–64.

Dempster, A., Laird, N., & Rubin, D. (1977). Maximum likelihood from incomplete data via the EM algorithm. *Journal of the Royal Statistical Society, Series B (Methodological)*, 39, 1–38.

Derwing, B. L., & Skousen, R. (1989). Real time morphology: Symbolic rules or analogical networks? *Berkeley Linguistic Society, 15,* 48–62.

Dietterich, T. (1998). Approximate Statistical Tests for Comparing Supervised Classification Learning Algorithms. *Neural Computation, 10,* 1895–1924.

Dietterich, T. G., Hild, H., & Bakiri, G. (1995). A comparison of ID3 and backpropagation for English text-to-speech mapping. *Machine Learning, 19,* 5–28.

Domingos, P. (1995). *The RISE 2.0 system: A case study in multistrategy learning* (Technical Report 95-2). University of California at Irvine, Department of Information and Computer Science, Irvine, CA.

Domingos, P. (1996). Unifying instance-based and rule-based induction. *Machine Learning, 24,* 141–168.

Dudani, S. (1976). The distance-weighted *k*-nearest neighbor rule. *IEEE Transactions on Systems, Man, and Cybernetics,* 325–327.

Durieux, G., & Gillis, S. (2000). Predicting grammatical classes from phonological cues: An empirical test. In B. Höhle and J. Weissenborn (Eds.), *Approaches to bootstrapping: Phonological, syntactic and neurophysiological aspects of early language acquisition,* 189–232. Amsterdam: Benjamins.

Eddington, D. (2003). Issues in modeling language processing analogically. *Lingua, 114,* 849–871.

Egan, J. P. (1975). *Signal detection theory and ROC analysis.* Series in Cognition and Perception. New York, NY: Academic Press.

Eisner, J. (1996). *An empirical comparison of probability models for dependency grammar.* Technical Report IRCS-96-11, Institute for Research in Cognitive Science, University of Pennsylvania.

Elman, J. (1990). Finding structure in time. *Cognitive Science, 14,* 179–211.

Estes, W. K. (1994). *Classification and cognition,* vol. 22 of *Oxford Psychology Series.* New York: Oxford University Press.

Evans, R. (2001). Applying machine learning toward an automatic classification of it. *Journal of Literary and Linguistic Computing, 16,* 45–57.

Fawcett, T. (2004). *ROC graphs: Notes and practical considerations for researchers* (Technical Report HPL-2003-4). Hewlett Packard Labs.

Fernández, R., Ginzburg, J., & Lappin, S. (2004). Classifying ellipsis in dialogue: A machine learning approach. *Proceedings of the 20th International Conference on Computational Linguistics, COLING 2004,* 240–246. Geneva, Switzerland.

Fillmore, C., Johnson, C., & Petruck, M. (2003). Background to framenet. *International Journal of Lexicography, 16,* 235–250.

BIBLIOGRAPHY

Fix, E., & Hodges, J. L. (1951). *Disciminatory analysis—nonparametric discrimination; consistency properties* (Technical Report Project 21-49-004, Report No. 4). USAF School of Aviation Medicine.

Fix, E., & Hodges, J. L. (1952). *Discriminatory analysis: Small sample performance* (Technical Report Project 21-49-004, Report No. unknown). USAF School of Aviation Medicine.

Franz, A. (1996). Learning PP attachment from corpus statistics. In S. Wermter, E. Riloff and G. Scheler (Eds.), *Connectionist, statistical, and symbolic approaches to learning for natural language processing*, vol. 1040 of *Lecture Notes in Artificial Intelligence*, 188–202. New York: Springer-Verlag.

Frazier, L. (1979). *On comprehending sentences: Syntactic parsing strategies*. Doctoral dissertation, University of Connecticut.

Frazier, L., & Clifton, C. (1998). *Construal*. Cambridge, MA: MIT Press.

Gazdar, G., & Mellish, C. (1989). *Natural language processing in LISP*. Reading, MA: Addison-Wesley.

Gillis, S., Durieux, G., & Daelemans, W. (2000). Lazy learning: A comparison of natural and machine learning of stress. *Cognitive Models of Language Acquisition*, 76–99. Cambridge University Press.

Glushko, R. J. (1979). The organisation and activation of orthographic knowledge in reading aloud. *Journal of Experimental Psychology: Human Perception and Performance*, 5, 647–691.

Halliday, M. A. K. (1961). Categories of the theory of grammar. *Word*, 17, 241–292.

Hammerton, J., Osborne, M., Armstrong, S., & Daelemans, W. (2002). *Special issue of journal of machine learning research on shallow parsing*. The MIT Press.

Harris, Z. S. (1940). Review of Louis H. Gray, Foundations of Language (New York: Macmillan, 1939). *Language*, 16, 216–231. Page numbers cited from repr. in Harris 1970:695–705 under the title "Gray's Foundations of Language".

Harris, Z. S. (1951). *Methods in structural linguistics*. University of Chicago Press.

Harris, Z. S. (1957). Co-occurrence and transformation in linguistic structure. *Language*, 33, 283–340.

Harris, Z. S. (1970). *Papers in structural and transformational linguistics*. No. 1 in Formal Linguistic Series. D. Reidel.

Hart, P. E. (1968). The condensed nearest neighbor rule. *IEEE Transactions on Information Theory*, 14, 515–516.

Heemskerk, J. (1993). A probabilistic context-free grammar for disambiguation in morphological parsing. *Proceedings of the 6th Conference of the EACL*, 183–192.

Heemskerk, J., & Van Heuven, V. J. (1993). Morpa, a lexicon-based morphological parser. In V. J. Van Heuven and L. C. W. Pols (Eds.), *Analysis and synthesis of speech; strategic research towards high-quality text-to-speech generation*. Berlin, Mouton de Gruyter.

Hendrickx, I., & Van den Bosch, A. (2003). Memory-based one-step named-entity recognition: Effects of seed list features, classifier stacking, and unannotated data. *Proceedings of CoNLL-2003*, 176–179.

Hindle, D., & Rooth, M. (1993). Structural ambiguity and lexical relations. *Computational Linguistics, 19,* 103–120.

Hoste, V. (2005). *Optimization in machine learning of coreference resolution*. Doctoral dissertation, University of Antwerp.

Hoste, V., Hendrickx, I., Daelemans, W., & Van den Bosch, A. (2002). Parameter optimization for machine learning of word sense disambiguation. *Natural Language Engineering, 8,* 311–325.

Jijkoun, V., & de Rijke, M. (2004). Enriching the output of a parser using memory-based learning. *Proceedings of the 42nd Meeting of the Association for Computational Linguistics (ACL'04), Main Volume,* 311–318. Barcelona, Spain.

Johansen, M., & Palmeri, T. (2002). Are there representational shifts during category learning? *Cognitive Psychology, 45,* 482–553.

Johnstone, T., & Shanks, D. R. (2001). Abstractionist and processing accounts of implicit learning. *Cognitive Psychology, 42,* 61–112.

Jones, D. (1996). *Analogical natural language processing*. London, UK: UCL Press.

Jurafsky, D., & Martin, J. H. (2000). *Speech and language processing: An introduction to natural language processing, computational linguistics, and speech recognition*. Englewood Cliffs, New Jersey: Prentice Hall.

Kasif, S., Salzberg, S., Waltz, D., Rachlin, J., & Aha, D. K. (1998). A probabilistic framework for memory-based reasoning. *Artificial Intelligence, 104,* 287–311.

Katz, S. M. (1987). Estimation of probabilities from sparse data for the language model component of a speech recognizer. *IEEE Transactions on Acoustics, Speech and Signal Processing, ASSP-35,* 400–401.

Kingsbury, P., Palmer, M., & Marcus, M. (2002). Adding semantic annotation to the Penn Treebank. *Proceedings of the Human Language Technology Conference*. San Diego, CA.

Knuth, D. E. (1973). *The art of computer programming*, vol. 3: Sorting and searching. Reading, MA: Addison-Wesley.

Kohavi, R., & John, G. (1997). Wrappers for feature subset selection. *Artificial Intelligence Journal, 97,* 273–324.

Kohonen, T. (1986). Dynamically expanding context, with application to the correction of symbol strings in the recognition of continuous speech. *Proceedings of the Eighth International Conference on Pattern Recognition*, 27–31. Paris, France.

Kokkinakis, D. (2000). PP-attachment disambiguation for swedish: Combining unsupervised and supervised training data. *Nordic Journal of Linguistics, 23*, 191–213.

Kolodner, J. (1993). *Case-based reasoning*. San Mateo, CA: Morgan Kaufmann.

Koskenniemi, K. (1983). Two-level model for morphological analysis. *Proceedings of the 8th International Joint Conference on Artificial Intelligence*. Los Alamos, CA: Morgan Kaufmann.

Koskenniemi, K. (1984). A general computational model for wordform recognition and production. *Proceedings of the Tenth International Conference on Computational Linguistics / 22nd Annual Conference of the Association for Computational Linguistics*, 178–181.

Krott, A., Baayen, R. H., & Schreuder, R. (2001). Analogy in morphology: modeling the choice of linking morphemes in Dutch. *Linguistics, 39*, 51–93.

Kübler, S. (2004). *Memory-based parsing*. Amsterdam, The Netherlands: John Benjamins.

Lafferty, J., McCallum, A., & Pereira, F. (2001). Conditional random fields: Probabilistic models for segmenting and labeling sequence data. *Proceedings of the 18th International Conference on Machine Learning*. Williamstown, MA.

Langley, P. (1996). *Elements of machine learning*. San Mateo, CA: Morgan Kaufmann.

Lavrac, N., & Džeroski, S. (1994). *Inductive logic programming*. Chichester, UK: Ellis Horwood.

Lendvai, P., Van den Bosch, A., & Krahmer, E. (2003a). Machine learning for shallow interpretation of user utterances in spoken dialogue systems. *Proceedings of the EACL Workshop on Dialogue Systems: Interaction, adaptation and styles of management*, 69–78.

Lendvai, P., Van den Bosch, A., & Krahmer, E. (2003b). Memory-based disfluency chunking. *Proceedings of Disfluency in Spontaneous Speech Workshop (DISS'03)*, 63–66.

Lendvai, P., Van den Bosch, A., Krahmer, E., & Swerts, M. (2002). Improving machine-learned detection of miscommunications in human-machine dialogues through informed data splitting. *Proceedings of the ESSLLI Workshop on Machine Learning Approaches in Computational Linguistics*.

Ling, C. X., & Wang, H. (1996). A decision-tree model for reading aloud with automatic alignment and grapheme generation. Submitted.

Lucassen, J. M., & Mercer, R. L. (1984). An information theoretic approach to the automatic determination of phonemic baseforms. *Proceedings of* ICASSP *'84, San Diego*, 42.5.1–42.5.4.

Luk, R., & Damper, R. (1996). Stochastic phonographic transduction for English. *Computer Speech and Language*, 10, 133–153.

MacLeod, J. E. S., Luk, A., & Titterington, D. M. (1987). A re-examination of the distance-weighted k-nearest neighbor classification rule. *IEEE Transactions on Systems, Man, and Cybernetics*, SMC-17, 689–696.

Malouf, R. (2000). The order of prenominal adjectives in natural language generation. *Proceedings of the 38th Annual Meeting of the Association for Computational Linguistics*, 85–92. New Brunswick, NJ: ACL.

Manning, C., & Schütze, H. (1999). *Foundations of statistical natural language processing*. Cambridge, Massachusetts: The MIT Press.

Marcus, G., Brinkmann, U., Clahsen, H., Wiese, R., & Pinker, S. (1995). German inflection: The exception that proves the rule. *Cognitive Psychology*, 29, 189–256.

Marcus, M., Santorini, S., & Marcinkiewicz, M. (1993). Building a Large Annotated Corpus of English: the Penn Treebank. *Computational Linguistics*, 19, 313–330.

Marsi, E., Reynaert, M., Van den Bosch, A., Daelemans, W., & Hoste, V. (2003). Learning to predict pitch accents and prosodic boundaries in Dutch. *Proceedings of the 41st Annual Meeting of the Association for Computational Linguistics*, 489–496. New Brunswick, NJ: ACL.

McCallum, A., Freitag, D., & Pereira, F. (2000). Maximum entropy Markov models for information extraction and segmentation. *Proceedings of the 17th International Conference on Machine Learning*. Stanford, CA.

Michalski, R. S. (1983). A theory and methodology of inductive learning. *Artificial Intelligence*, 11, 111–161.

Mihalcea, R. (2002). Instance-based learning with automatic feature selection applied to word sense disambiguation. *Proceedings of the 19th International Conference on Computational Linguistics (COLING 2002)*. Taipei, Taiwan.

Minnen, G., Bond, F., & Copestake, A. (2000). Memory-based learning for article generation. *Proceedings of the 4th Conference on Computational Natural Language Learning and the Second Learning Language in Logic Workshop*, 43–48. New Brunswick, NJ: ACL.

Mitchell, T. (1997). *Machine learning*. New York, NY: McGraw-Hill.

Mitkov, R. (Ed.). (2003). *The Oxford Handbook of Computational Linguistics*. Oxford: Oxford University Press.

Mitkov, R., Evans, R., & Orasan, C. (2002). A new, fully automatic version of Mitkov's knowledge-poor pronoun resolution method. *Proceedings of the Third International Conference on Computational Linguistics and Intelligent Text Processing*, 168–186. Springer-Verlag.

Morin, R. L., & Raeside, B. E. (1981). A reappraisal of distance-weighted k-nearest neighbor classification for pattern recognition with missing data. *IEEE Transactions on Systems, Man, and Cybernetics, SMC-11*, 241–243.

Nagao, M. (1984). A framework of a mechanical translation between Japanese and English by analogy principle. In A. Elithorn and R. Banerji (Eds.), *Artificial and human intelligence*, 173–180. Amsterdam: North-Holland.

Niblett, T. (1987). Constructing decision trees in noisy domains. *Proceedings of the Second European Working Session on Learning*, 67–78. Bled, Yugoslavia: Sigma.

Nivre, J., Hall, J., & Nilsson, J. (2004). Memory-based dependency parsing. *Proceedings of the Eighth Conference on Computational Natural Language Learning (CoNLL 2004)*, 49–57. Boston, Massachusetts.

Nivre, J., & Scholz, M. (2004). Deterministic dependency parsing of English text. *Proceedings of COLING 2004*, 23–27. Geneva, Switzerland.

Nosofsky, R. (1986). Attention, similarity, and the identification-categorization relationship. *Journal of Experimental Psychology: General, 15*, 39–57.

Okabe, A., Boots, B., Sugihara, K., & Chiu, S. N. (2000). *Spatial tesselations: Concepts and applications of Voronoi diagrams*. John Wiley. Second edition.

Orăsan, C., & Evans, R. (2001). Learning to identify animate references. *Proceedings of the Fifth Workshop on Computational Language Learning, CoNLL-2001*, 129–136. Toulouse, France.

Palmer, F. R. (Ed.). (1969). *Selected papers of J. R. Firth 1952–1959*. London: Longmans.

Piatelli–Palmarini, M. (Ed.). (1980). *Language learning: The debate between Jean Piaget and Noam Chomsky*. Cambridge, MA: Harvard University Press.

Pirelli, V., & Federici, S. (1994). On the pronunciation of unknown words by analogy in text-to-speech systems. *Proceedings of the Second Onomastica Research Colloquium*. London.

Pohlmann, R., & Kraaij, W. (1997). Improving the precision of a text retrieval system with compound analysis. *CLIN VII – Papers from the Seventh CLIN meeting*, 115–128.

Preiss, J. (2002). Anaphora resolution with memory-based learning. *Proceedings of the Fifth Annual CLUK Research Colloquium*, 1–9.

Provost, F., Jensen, D., & Oates, T. (1999). Efficient progressive sampling. *Proceedings of the Fifth International Conference on Knowledge Discovery and Data Mining*, 23–32.

Quinlan, J. (1986). Induction of Decision Trees. *Machine Learning*, 1, 81–206.

Quinlan, J. (1993). C4.5: *Programs for machine learning*. San Mateo, CA: Morgan Kaufmann.

Ramshaw, L., & Marcus, M. (1995). Text chunking using transformation-based learning. *Proceedings of the 3rd ACL/SIGDAT Workshop on Very Large Corpora, Cambridge, Massachusetts, USA*, 82–94.

Ratnaparkhi, A. (1996). A maximum entropy part-of-speech tagger. *Proceedings of the Conference on Empirical Methods in Natural Language Processing, May 17-18, 1996, University of Pennsylvania*.

Ratnaparkhi, A. (1998). *Maximum entropy models for natural language ambiguity resolution*. Doctoral dissertation, University of Pennsylvania.

Ratnaparkhi, A., Reynar, J., & Roukos, S. (1994). A maximum entropy model for prepositional phrase attachment. *Workshop on Human Language Technology*. Plainsboro, NJ.

Reinberger, M.-L., Spyns, P., Pretorius, A. J., & Daelemans, W. (2004). Automatic initiation of an ontology. *On the Move to Meaningful Internet Systems 2004: CoopIS, DOA, and ODBASE, OTM Confederated International Conferences*, 600–617.

Riesbeck, C., & Schank, R. (1989). *Inside case-based reasoning*. Northvale, NJ: Erlbaum.

Rissanen, J. (1983). A universal prior for integers and estimation by minimum description length. *Annals of Statistics*, 11, 416–431.

Rotaru, M., & Litman, D. (2003). Exceptionality and natural language learning. *Proceedings of the Seventh Conference on Computational Natural Language Learning*. Edmonton, Canada.

Roth, D. (1998). Learning to resolve natural language ambiguities: A unified approach. *Proceedings of the National Conference on Artificial Intelligence*, 806–813. Menlo Park, CA: AAAI Press.

Salzberg, S. (1990). *Learning with nested generalised exemplars*. Norwell, MA: Kluwer Academic Publishers.

Salzberg, S. (1991). A nearest hyperrectangle learning method. *Machine Learning*, 6, 277–309.

Salzberg, S. (1997). On comparing classifiers: Pitfalls to avoid and a recommended approach. *Data Mining and Knowledge Discovery*, 1.

Scha, R. (1992). Virtual Grammars and Creative Algorithms. *Gramma/TTT Tijdschrift voor Taalkunde*, 1, 57–77.

Scha, R., Bod, R., & Sima'an, K. (1999). A memory-based model of syntactic analysis: data-oriented parsing. *Journal of Experimental and Theoretical Artificial Intelligence*, 11, 409–440.

Schölkopf, B., Burges, C., & Vapnik, V. (1995). Extracting support data for a given task. *Proceedings of the First International Conference on Knowledge Discovery and Data Mining.* Menlo Park: AAAI Press.

Sejnowski, T., & Rosenberg, C. (1987). Parallel networks that learn to pronounce English text. *Complex Systems, 1,* 145–168.

Sejnowski, T. J., & Rosenberg, C. (1986). *NETtalk:A parallel network that learns to read aloud* (Technical Report JHU EECS 86-01). Johns Hopkins University.

Shepard, R. (1987). Toward a universal law of generalization for psychological science. *Science, 237,* 1317–1323.

Skousen, R. (1989). *Analogical modeling of language.* Dordrecht: Kluwer Academic Publishers.

Skousen, R. (2002). An overview of analogical modeling. In R. Skousen, D. Lonsdale and D. B. Parkinson (Eds.), *Analogical modeling: An exemplar-based approach to language,* 11–26. Amsterdam, The Netherlands: John Benjamins.

Skousen, R., Lonsdale, D., & Parkinson, D. B. (Eds.). (2002). *Analogical modeling: An exemplar-based approach to language.* Amsterdam, The Netherlands: John Benjamins.

Smith, E., & Medin, D. (1981). *Categories and concepts.* Cambridge, MA: Harvard University Press.

Smith, L., & Samuelson, L. (1997). Perceiving and remembering: Category stability, variability, and development. In K. Lamberts and D. Shanks (Eds.), *Knowledge, concepts, and categories,* 161–195. Cambridge: Cambridge University Press.

Spitters, M. (2000). Comparing feature sets for learning text categorization. *Proceedings of the Sixth Conference on Content-Based Multimedia Access (RIAO 2002),* 1124–1135. Paris, France.

Sproat, R. (1992). *Morphology and computation.* ACL-MIT Press Series in Natural Language Processing. Cambridge, MA: The MIT Press.

Stanfill, C. (1987). Memory-based reasoning applied to English pronunciation. *Proceedings of the Sixth National Conference on Artificial Intelligence,* 577–581. Los Altos, CA: Morgan Kaufmann.

Stanfill, C., & Waltz, D. (1986). Toward memory-based reasoning. *Communications of the ACM, 29,* 1213–1228.

Stetina, J., & Nagao, M. (1997). Corpus-based PP attachment ambiguity resolution with a semantic dictionary. *Proceedings of the Fifth Workshop on Very Large Corpora,* 66–80. Beijing, China.

Stevenson, M., & Wilks, Y. (1999). Combining weak knowledge sources for sense disambiguation. *Proceedings of the International Joint Conference on Artificial Intelligence.*

Streiter, O. (2001a). Memory-based parsing: Enhancing recursive top-down fuzzy match with bottom-up chunking. *Proceedings of the Nineteenth International Conference on Computer Processing of Oriental Languages (ICCPOS 2001).*

Streiter, O. (2001b). Recursive top-down fuzzy match: New perspectives on memory-based parsing. *Proceedings of the Fifteenth Pacific Asia conference on Language, Information and Computation (PACLIC 2001).*

Sullivan, K., & Damper, R. (1992). Novel-word pronunciation with a text-to-speech system. In G. Bailly and C. Benoît (Eds.), *Talking machines: theories, models, and applications,* 183–195. Amsterdam: Elsevier.

Sullivan, K., & Damper, R. (1993). Novel-word pronunciation: a cross-language study. *Speech Communication, 13,* 441–452.

Sun, R., & Giles, L. (2001). *Sequence learning: Paradigms, algorithms, and applications.* Heidelberg: Springer Verlag.

Swets, J., Dawes, R., & Monahan, J. (2000). Better decisions through science. *Scientific American, 283,* 82–87.

Swonger, C. W. (1972). Sample set condensation for a condensed nearest neighbor decision rule for pattern recognition. In S. Watanabe (Ed.), *Frontiers of pattern recognition,* 511–519. Orlando, Fla: Academic Press.

Thompson, C. A., Califf, M. E., & Mooney, R. J. (1999). Active learning for natural language parsing and information extraction. *Proceedings of the Sixteenth International Conference on Machine Learning,* 406–414. Morgan Kaufmann, San Francisco, CA.

Tjong Kim Sang, E. (2002). Memory-based shallow parsing. *Journal of Machine Learning Research, 2,* 559–594.

Tjong Kim Sang, E., & Buchholz, S. (2000). Introduction to the CoNLL-2000 shared task: Chunking. *Proceedings of CoNLL-2000 and LLL-2000,* 127–132.

Tjong Kim Sang, E., & De Meulder, F. (2003). Introduction to the CoNLL-2003 shared task: Language-independent named entity recognition. *Proceedings of CoNLL-2003,* 142–147. Edmonton, Canada.

Tjong Kim Sang, E., & Veenstra, J. (1999). Representing text chunks. *Proceedings of EACL'99,* 173–179. Bergen, Norway.

Tomasello, M. (2003). *Constructing a language: A usage-based theory of language acquisition.* Harvard University Press.

Tomek, I. (1976). An experiment with the edited nearest-neighbor rule. *IEEE Transactions on Systems, Man, and Cybernetics, SMC-6,* 448–452.

Torkkola, K. (1993). An efficient way to learn English grapheme-to-phoneme rules automatically. *Proceedings of the International Conference on Acoustics, Speech, and Signal Processing (ICASSP),* 199–202. Minneapolis.

Twain, M. (1880). *A tramp abroad*. Hartford: American Publishing Co.

Van den Bosch, A. (1997). *Learning to pronounce written words: A study in inductive language learning*. Doctoral dissertation, Universiteit Maastricht.

Van den Bosch, A. (1999). Instance-family abstraction in memory-based language learning. *Machine Learning: Proceedings of the Sixteenth International Conference*, 39–48. Bled, Slovenia.

Van den Bosch, A., & Buchholz, S. (2002). Shallow parsing on the basis of words only: A case study. *Proceedings of the 40th Meeting of the Association for Computational Linguistics*, 433–440.

Van den Bosch, A., Canisius, S., Daelemans, W., Hendrickx, I., & Tjong Kim Sang, E. (2004). Memory-based semantic role labeling: Optimizing features, algorithm, and output. *Proceedings of the Eighth Conference on Computational Natural Language Learning*. Boston, MA, USA.

Van den Bosch, A., Content, A., Daelemans, W., & De Gelder, B. (1995). Measuring the complexity of writing systems. *Journal of Quantitative Linguistics, 1*.

Van den Bosch, A., & Daelemans, W. (1993). Data-oriented methods for grapheme-to-phoneme conversion. *Proceedings of the 6th Conference of the EACL*, 45–53.

Van den Bosch, A., & Daelemans, W. (1999). Memory-based morphological analysis. *Proceedings of the 37th Annual Meeting of the ACL*, 285–292. San Francisco, CA: Morgan Kaufmann.

Van den Bosch, A., Daelemans, W., & Weijters, A. (1996). Morphological analysis as classification: an inductive-learning approach. *Proceedings of the Second International Conference on New Methods in Natural Language Processing, NeMLaP-2, Ankara, Turkey*, 79–89.

Van den Bosch, A., Krahmer, E., & Swerts, M. (2001). Detecting problematic turns in human-machine interactions: Rule-induction versus memory-based learning approaches. *Proceedings of the 39th Meeting of the Association for Computational Linguistics*, 499–506. New Brunswick, NJ: ACL.

Van Halteren, H. (1999). *Syntactic wordclass tagging*. Dordrecht, The Netherlands: Kluwer Academic Publishers.

Van Halteren, H., Zavrel, J., & Daelemans, W. (2001). Improving accuracy in word class tagging through combination of machine learning systems. *Computational Linguistics, 27*, 199–230.

Van Herwijnen, O., Van den Bosch, A., Terken, J., & Marsi, E. (2004). Learning PP attachment for filtering prosodic phrasing. *Tenth Conference of the European Chapter of the Association for Computational Linguistics (EACL-03)*, 139–146.

Van Rijsbergen, C. (1979). *Information retrieval*. London: Buttersworth.

Vapnik, V., & Bottou, L. (1993). Local algorithms for pattern recognition and dependencies estimation. *Neural Computation, 5*, 893–909.

Varges, S., & Mellish, C. (2001). Instance-based natural language generation. *Proceedings of the 2nd Meeting of the North American Chapter of the Association for Computational Linguistics (NAACL-01)*, 1–8. New Brunswick, NJ: ACL.

Veenstra, J. (1998). Fast NP chunking using memory-based learning techniques. *Proceedings of BENELEARN'98*, 71–78. Wageningen, The Netherlands.

Veenstra, J., & Daelemans, W. (2000). *A memory-based alternative for connectionist shift-reduce parsing* (Technical Report ILK 00-12). ILK Research Group, University of Tilburg.

Veenstra, J., Van den Bosch, A., Buchholz, S., Daelemans, W., & Zavrel, J. (2000). Memory-based word sense disambiguation. *Computers and the Humanities, 34*, 171–177.

Viterbi, A. J. (1967). Error bounds for convolutional codes and an asymptotically optimum decoding algorithm. *IEEE Transactions on Information Theory, 13*, 260–269.

Weijters, A. (1991). A simple look-up procedure superior to NETtalk? *Proceedings of the International Conference on Artificial Neural Networks - ICANN-91*, Espoo, Finland.

Weiss, S., & Kulikowski, C. (1991). *Computer systems that learn*. San Mateo, CA: Morgan Kaufmann.

Wermter, S., Riloff, E., & Scheler, G. (1996). Learning aprroaches for natural language processing. In S. Wermter, E. Riloff and G. Scheler (Eds.), *Connectionist, statistical and symbolic approaches to learning for natural language processing*, vol. 1040 of *Lecture Notes in Artificial Intelligence*, 1–16. Berlin: Springer.

Wettschereck, D. (1994). *A study of distance-based machine learning algorithms*. Doctoral dissertation, Oregon State University.

Wettschereck, D., & Dietterich, T. G. (1994). Locally adaptive nearest neighbor algorithms. *Advances in Neural Information Processing Systems*, 184–191. Palo Alto, CA: Morgan Kaufmann.

Wettschereck, D., & Dietterich, T. G. (1995). An experimental comparison of the nearest-neighbor and nearest-hyperrectangle algorithms. *Machine Learning, 19*, 1–25.

White, A., & Liu, W. (1994). Bias in information-based measures in decision tree induction. *Machine Learning, 15(3)*, 321–329.

Wilson, D. (1972). Asymptotic properties of nearest neighbor rules using edited data. *Institute of Electrical and Electronic Engineers Transactions on Systems, Man and Cybernetics, 2*, 408–421.

Wilson, D., & Martinez, A. (1997). Instance pruning techniques. *Machine Learning: Proceedings of the Fourteenth International Conference*. San Francisco, CA.

Wolpert, D. H. (1992). Stacked Generalization. *Neural Networks*, 5, 241–259.

Yvon, F. (1996). *Prononcer par analogie: motivation, formalisation et évaluation*. Doctoral dissertation, Ecole Nationale Supérieure des Télécommunications, Paris.

Zavrel, J. (1997). An empirical re-examination of weighted voting for k-NN. *Proceedings of the 7th Belgian-Dutch Conference on Machine Learning*, 139–148. Tilburg.

Zavrel, J., Berck, P., & Lavrijssen, W. (2000). Information extraction by text classification: Corpus mining for features. *Proceedings of the workshop Information Extraction meets Corpus Linguistics*. Athens, Greece.

Zavrel, J., & Daelemans, W. (1997). Memory-based learning: Using similarity for smoothing. *Proceedings of the 35th Annual Meeting of the Association for Computational Linguistics*, 436–443.

Zavrel, J., & Daelemans, W. (1999). Recent advances in memory-based part-of-speech tagging. *VI Simposio Internacional de Comunicacion Social, Santiago de Cuba*, 590–597.

Zavrel, J., & Daelemans, W. (2003). Feature-rich memory-based classification for shallow NLP and information extraction. *Text Mining, Theoretical Aspects and Applications*, 33–54. Heidelberg, Germany: Springer Physica-Verlag.

Zavrel, J., Daelemans, W., & Veenstra, J. (1997). Resolving PP attachment ambiguities with memory-based learning. *Proceedings of the Workshop on Computational Language Learning (CoNLL'97)*. ACL, Madrid.

Zhang, J. (1992). Selecting typical instances in instance-based learning. *Proceedings of the International Machine Learning Conference 1992*, 470–479.

Index

abstraction, 24, 104–147
ACL, 14
affix, 57
ambitag, 87
analogical modeling, 18
analogy, 15, 16
anaphora resolution, 166
artificial intelligence, 5
attenuation, 88
AUC, area under the curve, 50
automatic subtitling, 102

back-off estimation, 146
Bloomfield, L., 15
BNGE, 129
Brill tagger, 109

C4.5, 64, 83, 104, 129, 144, 145
case-based reasoning, 22
CELEX, 28, 58, 69, 74, 81, 108, 144
χ^2 statistic, 31
CHUNK, see English constituent chunking
chunking, 96
city-block distance, 29
class n-grams, 160–161
class prediction strength, 116–123
classification, 16
clause boundary detection, 102
CN2, 130
cognitive linguistics, 19, 20
compounding, 74

computational learning theory, 104
computational morphology, 84
conditional random fields, 156
confusion matrix, 48, 49
CoNLL, 14, 109
constituent chunking, 96, 109
core, 123

data-oriented parsing, 24, 146
De Saussure, F., 15
decision tree induction, 5, 63
derivation, 74
dialog
    act classification, 166
    systems, 146, 166
DIMIN, see Dutch diminutive formation
diminutive formation, 107
Dirichlet tile, 117
disambiguation, 8
disfluencies, 166
disjunctivity, 123
distance weighting, 42–44
    exponential decay, 43
    inverse, 43
    inverse-linear, 42
DOP, data-oriented parsing, 24, 103
dual route model, 27
Dutch
    compound linking, 84
    diminutive formation, 107
    morphological analysis, 74, 108

INDEX

morphology, 74, 84
prosody generation, 166
word phonemization, 84
word stress, 167

EACH, 116
eager learning, 22
ED, exponential-decay distance weighting, 43
editing, 115–127, 146
elsewhere condition, 123
EMNLP, 14
English
    constituent chunking, 96, 109
    named entity recognition, 109
    past tense, 84
    prepositional-phrase attachment, 9, 108
entropy, 30
example-based machine translation, 23
exemplar-based learning, 22
expectation-maximization, 60

F-score, 50, 78
FAMBL, 132–143, 145
feature weighting, 29–32
    χ^2, 31
    gain ratio, 30
    information gain, 29
    shared variance, 32
finite-state transducers, 84
Firth, J. R., 17
FPR, false positive rate, 49
French
    word phonemization, 84

gain ratio, 30
generalization, 24, 104–147
German
    plural formation, 27, 107

GPLURAL, *see* German plural formation
GR, gain ratio, 30
grammatical relations, 99

Halliday, M. A. K., 17
Hamming distance, 29
harmonic mean, 50
Harris, Z., 18
Hidden Markov models, 164
hyperplane, 123
hyperrectangle, 129

IB1, 29
IB2, 144
IB3, 116, 144
ICF, 146
ID, inverse distance weighting, 43
ID3, 64
IG, information gain, 29
IGTREE, 64–71, 111, 129, 145
IL, inverse-linear distance weighting, 42
induction, 17
inductive logic programming, 5
inflection, 74
information extraction, 85, 167
information gain, 29
information retrieval, 81, 85, 87
information theory, 29
instance base, 26
instance-based learning, 22
IOB tags, 96, 109, 110

k–d trees, 21
k–nearest neighbor classifier, 21, 29

L1 metric, 29
language
    acquisition, 167
    experience-based, 20

engineering, 3
signs, 16
technology, 3
langue, 16
Laplace correction, 117, 131
late closure principle, 11
lazy learning, 22
learning curve, 53
leave-one-out validation, 47
lexical lookup, 58
lexicon, 58
locally-weighted learning, 22

machine learning, 5
machine translation
    example-based, 23
MALT parser, 103
Manhattan distance, 29
maximum-entropy markov models, 156
maximum-likelihood estimation, 146
MBRTalk, 83
MBT, 90
MBTG, 90
memory-based reasoning, 22
memory-based sequence learning, 103
minimal attachment principle, 11
minimal description length principle, 5, 104, 143
MITALK, 59
modified value difference metric, 38–41
MORPH, *see* Dutch morphological analysis
morpheme, 57
morphological analysis, 73–80
morphology, 57

named-entity recognition, 109, 167
natural language processing, 3
nearest-neighbor classifier, 21
NER, *see* English named entity recognition
Nettalk, 83
NeXTeNS, 81
NGE, 129
noun
    animacy, 166
    countability, 166
    plural formation, 27

Ockham's razor, 5, 104
OCTOPUS parser, 103
ontology extraction, 85, 102
optimality theory, 167
orthography, 59, 66
overfitting, 149
overlap metric, 29

paired t-test, 55
paradigmatic relation, 16
paramsearch, 152
parole, 16
parsing, 85, 87
part-of-speech tagging, 86
partial parsing, 85
Penn Treebank, 108, 109, 144
periphery, 123
phoneme, 57
phonemization, 59–73
phonology, 57
plural formation, 107
polymorphism, 124
POS, 86
PP, *see* English prepositional-phrase attachment
precision, 49
principles and parameters, 167

INDEX

progressive sampling, 150
prosody, 166
psychology, 19

question answering, 85, 102

recall, 49
receiver operator characteristics, 50
recurrent neural networks, 156
relation finding, 86, 99
Reuters Corpus RCV1, 109
RIPPER, 104–111, 144–146
RISE, 129, 130
ROC space, 50
rule induction, 5, 106

segmentation, 8, 16
SENSEVAL, 166
sentence generation, 167
shallow parsing, 85–103
shared variance, 32
shift-reduce parsing, 103
SIGDAT, 14
SIGNLL, 14, 102
similarity metric, 28–44
Skousen, R., 18
spam filtering, 167
Spanish
    word stress assignment, 84
speech synthesis, 59, 87
spelling error correction, 87
spoken language parsing, 103
stacking, 157–160, 165
    adaptive, 159
    perfect, 158
statistical natural language processing, 17
statistical pattern recognition, 5, 21
stochastic grammars, 164
stochastic transducers, 84
substitution grammars, 18

summarization, 85, 102
support vector machines, 123, 144
syntagmatic relations, 16
systemic functional grammar, 17

ten-fold cross-validation, 47
text classification, 167
text generation, 166
text mining, 87, 167
TiMBL, 26, 32–41, 44, 45, 47, 48, 52, 67, 73
TPR, true positive rate, 49
transformation-based error-driven learning, 5
TreeTalk, 60
TRIBL, 71–73, 145
TüSBL, 103
two-level morphology, 84
typicality, 146

unknown words, 87
usage-based models of language, 20

Voronoi cell, 117

windowing, 9, 62, 156
word sense disambiguation, 166
word stress assignment, 167
wrapped progressive sampling, 149–155
wrapping, 149

计算语言学与语言科技原文丛书

INTRODUCING SPEECH AND LANGUAGE PROCESSING
语音语言处理导论
John Coleman　著　常宝宝　导读

LOGICS OF CONVERSATION
对话的逻辑
Nicholas Asher, Alex Lascarides　著　曾淑娟　导读

BUILDING NATURAL LANGUAGE GENERATION SYSTEMS
自然语言生成系统的建造
Ehud Reiter, Robert Dale　著　冯志伟　导读

THE SPOKEN LANGUAGE TRANSLATOR
口语机器翻译
Manny Rayner, David Carter, Pierrette Bouillon, Vassilis Digalakis, Mats Wirén　编
宗成庆　导读

THE LANGUAGE OF WORD MEANING
词义的语言
Pierrette Bouillon, Federica Busa　编　黄居仁　苏祺　导读

A COMPUTATIONAL THEORY OF WRITING SYSTEMS
文字书写系统的计算理论
Richard Sproat　著　陆勤　导读

WORD SENSE DISAMBIGUATION
词义消歧
Eneko Agirre, Philip Edmonds　编　赵铁军　导读

TREEBANKS
树库
Anne Abeillé　编　詹卫东　导读

WORD FREQUENCY DISTRIBUTIONS
词汇频率分布
R.Harald Baayen　著　张化瑞　导读

EVALUATION OF TEXT AND SPEECH SYSTEMS
文本和语音处理系统评测
Laila Dybkjær, Holmer Hemsen, Wolfgang Minker　编　徐飞玉　导读

ONTOLOGY AND THE LEXICON
本体与词汇库
Chu-Ren Huang, Nicoletta Calzolari, Aldo Gangemi,
Alessandro Lenci, Alessandro Oltramari, Laurent Prévot　编
陆勤　导读

MEMORY-BASED LANGUAGE PROCESSING
基于记忆的语言处理
Walter Daelemans, Antal van den Bosch　著　孙栩　导读